Oceanic Acidification
A Comprehensive Overview

Oceanic Acidification
A Comprehensive Overview

Ronald Eisler

Senior Scientist (retired)
U.S. Geological Survey

CRC Press
Taylor & Francis Group
Boca Raton London New York

CRC Press is an imprint of the
Taylor & Francis Group, an **informa** business

Science Publishers
Jersey, British Isles
Enfield, New Hampshire

CRC Press
Taylor & Francis Group
6000 Broken Sound Parkway NW, Suite 300
Boca Raton, FL 33487-2742

First issued in paperback 2017

ISBN-13: 978-1-57808-770-9 (hbk)
ISBN-13: 978-1-138-11805-8 (pbk)

Library of Congress Cataloging-in-Publication Data

Eisler, Ronald, 1932-
  Oceanic acidification : a comprehensive overview / Ronald
Eisler.
      p. cm.
  Includes bibliographical references and index.
  ISBN 978-1-57808-770-9 (hardcover)
  1. Ocean acidification--Environmental aspects. 2. Chemical
oceanography. I. Title.
  QH545.W3E37 2011
  551.46'6--dc23
                                              2011030376

Visit the Taylor & Francis Web site at
http://www.taylorandfrancis.com

and the CRC Press Web site at
http://www.crcpress.com

Preface

When we burn gasoline in our cars, burn natural gas or coal to produce electricity at power plants, or chop down and burn tropical forests for new agriculture, we release carbon dioxide (CO_2) gas into the atmosphere (Doney et al. 2009). The quantity of CO_2 released by human activities in 2008 alone exceeded 10 billion tons of carbon—equivalent to one million tons per hour or 0.2 kg per person per hour. Of this amount 8.7 billion tons originated from fossil fuel combustion and cement production and another 1.2 billion tons from deforestation (Doney et al. 2009). The 20th century increase in atmospheric CO_2 occurred more than ten times faster than any sustained change during the past 22,000 years, implying that global climate change—which is anthropogenic in origin—is progressing at a speed that is unprecedented during this period (Joos and Spahni 2008). Until recently most people thought it incredible that climate could change globally except on a geological timescale, pushed by forces far stronger than human activity (Weart 2010). In the late 1970s, a scientific consensus began to take shape, culminating around the year 2000 in unanimous agreement among governments on essential points, although many uncertainties remained. Increasing media warnings of peril made most of the literate world aware of the issue. However, skepticism persisted—correlated with aversion to regulation. At present, the majority of the world is now concerned about changes in global climate patterns, including potential oceanic acidification, but seems disinclined to take action (Weart 2010).

Literature Cited

Doney, S.C., W.M. Balch, V.J. Fabry, and R.E. Feely. 2009. Ocean acidification. A critical emerging problem for the ocean sciences, *Oceanography*, 22, 16–25.

Joos, F. and R. Spahni. 2008. Rates of change in natural and anthropogenic radiative forcing over the past 20,000 years, *Proc. Natl. Acad. Sci. USA*, 105, 1425–1430.

Weart, S.R. 2010. The ideas of anthropogenic global climate change in the 20th century, *Wiley Interdisciplinary Reviews: Climate Change*, 1, 67–81.

Acknowledgments

I am grateful to the librarians of the U.S. Department of Agriculture National Agricultural Library, located in Beltsville, Maryland, for providing most research materials.

Contents

Preface   v
Acknowledgments   vii

1. Introduction   **1**

1.1 General   1
1.2 Ocean Acidification   2
1.3 Terrestrial and Freshwater Acidification   5
1.4 Literature Cited   7

2. Historical   **11**

2.1 General   11
2.2 Present to 35,000 Years Ago   12
2.3 From 35,000 to 450,000 Years Ago   13
2.4 From 450,000 to 2.1 Million Years Ago   14
2.5 From 15 to 60 Million Years Ago   15
2.6 From 89 to 545+ Million Years Ago   16
2.7 Literature Cited   18

3. Sources of Oceanic Acidification   **21**

3.1 General   21
3.2 Anthropogenic   23
3.3 Biological   29
3.4 Physical   39
3.5 Literature Cited   42

4. Mode of Action   **49**

4.1 General   49
4.2 Chemical   52
4.3 Physical   57
4.4 Biological   59
4.5 Literature Cited   89

5. Acidification Effects on Biota   **99**

5.1 General   99
5.2 Photosynthetic Flora   102
5.3 Invertebrates   112
5.4 Vertebrates   140
5.5 Literature Cited   149

6. Field Studies   **164**

6.1 General   164
6.2 Arctic Ocean   165
6.3 Arabian Sea   166
6.4 Atlantic Ocean   166
6.5 Australia   170
6.6 Baltic Sea   171
6.7 Belgian Coastal Areas   171
6.8 Bering Sea and Environs   172
6.9 Bermuda   172
6.10 Borneo   173
6.11 Caribbean Region   173
6.12 Greenland Sea   174
6.13 Gulf of Maine   174
6.14 Indian Ocean   175
6.15 Ischia Island, Italy   176
6.16 Japan, Volcano Islands   177
6.17 Labrador Sea   177
6.18 North American West Coast   178
6.19 North Sea   178
6.20 Pacific Ocean   179
6.21 Red Sea   184
6.22 Southern Ocean   185
6.23 Tatoosh Island, Washington   187
6.24 Literature Cited   188

7. Modifiers   **193**

7.1 General   193
7.2 Methodological   193
7.3 Natural Variations   195
7.4 Interactions   199
7.5 Literature Cited   209

8. Mitigation   **213**

8.1 General   213
8.2 Ocean Sequestration   214
8.3 Declining Water Quality   218
8.4 Reduction in Emissions from Airliners   219
8.5 Increasing International Cooperation   219
8.6 Develop Alternative Technologies   222
8.7 Environmental Modification   222
8.8 Legislation   223
8.9 Literature Cited   224

9. Concluding Remarks   **228**

General Index   233
Species Index   245
About the Author   251

List of Tables

Table 1.1. Atmospheric CO_2 between the years 1000 and 2006   4

Table 3.1. Proposed anthropogenic CO_2 budget for the periods   25
1800–1994 and for 1980–1999

Table 3.2. Sediment trap particulate dissolution fluxes in the   27
Pacific Ocean

Table 7.1. Metal salts in surface seawater projected to undergo 204
at least a 50% change between the year 2000 (pH 8.1) and
2100 (pH 7.7)

Table 8.1. The top ten CO_2-emitting countries in 2006. Total   214
and per capita emissions

CHAPTER 1

Introduction

1.1 General

The ocean environment is expected to undergo significant changes in response to rising atmospheric carbon dioxide (CO_2) concentrations from the burning of fossil fuels such as oil, coal, and wood (Revelle and Suess 1957; Chen and Millero 1979). Systems affected include different ocean circulation patterns, changes in biotic diversity, altered ecosystems, a temperature increase of 1° to 3°C in seawater surface temperature, decreasing salinity from melting polar ice, and a predicted rise in sea level of 0.5 m in the next 50 to 100 years (as quoted in Palacios and Zimmerman 2007). Some scientists now believe that sea level rise by the year 2100 will be at least 1.0 m (Gillis 2010). Many scientists concur that atmospheric concentrations of greenhouse gases—especially carbon dioxide—are increasing, causing an increase in the average global temperature, resulting in a rise in sea level, extreme weather events, an increase in the spread of tropical diseases, species extinctions, and ecosystem changes (Karl and Trenberth 2003; Shellito et al. 2003; Portner et al. 2004; Bradshaw 2007; Millero 2007; Weart 2010). Ocean acidification, a consequence of anthropogenic carbon dioxide emissions, poses a serious threat to marine biota in tropical, open-ocean, coastal, deep-sea, and high latitude sea ecosystems (Hofmann et al. 2010).

Oceanic uptake of anthropogenic carbon dioxide may alter the seawater chemistry of the world's oceans with serious potential consequences for marine biota, including decreased calcification processes, increased acidity, phytoplankton declines, and reduced oxygen transport capacity (Fabry et al. 2008; Guinotte and Fabry

2008; Doney et al. 2009; Rijnsdorp et al. 2009; Boyce et al. 2010). Near-future ocean acidification will have dramatic negative impact on fishery and shellfish resources and other marine species, with likely consequences at the ecosystem level (Bradshaw 2007; Dupont et al. 2010; UNEP 2010). It is frequently assumed that these impacts will fade away soon thereafter, but recent models suggest that significant impacts will persist for hundreds of thousands of years after carbon dioxide emissions cease (Montenegro et al. 2007; Tyrrell et al. 2007). One challenge currently facing scientists is to predict the long term implications of ocean acidification on the diversity of marine organisms and on the ecosystem functions this diversity sustains (Widdicombe and Spicer 2008). The main goal of the present work is to critically assess the available information on changes in marine pH due to anthropogenic carbon dioxide generated from combustion of fossil fuels, and to assess the potential implications for marine life. Supporting chapters include: a historical perspective of CO_2 and ocean pH values through the millennia; biological, physical, and anthropogenic sources of ocean acidification; chemical, physical, and biological mode of acidification; projected acidification effects on marine flora, invertebrates, and vertebrates; results of field studies on carbon and its compounds in select marine bodies of water; physical and chemical modifiers known to affect pH and CO_2 values and effects; and proposed mitigation strategies including ocean sequestration, environmental legislation, and others.

1.2 Ocean Acidification

Between 1980 and 1989, the ocean stored 14.8 petagrams (1 Pg = 1 billion tons) of anthropogenic carbon and another 17.9 petagrams between 1990 and 1999, indicating an ocean wide net uptake of 1.6 and 2.0 petagrams per year, respectively (McNeil et al. 2003). During the past 200 years a number of greenhouse gases—including carbon dioxide, methane, nitrous oxide, and chlorofluorocarbons—have been increasing in the atmosphere, absorbing infrared energy resulting in an increase in the temperature of the troposphere. At the present time, the most important greenhouse gas is carbon dioxide (Haugan and Drange 1996; Raven et al. 2005; Millero 2007; Turley 2008; USDC 2008; Widdicombe and Spicer 2008). In the past 200 years the oceans have absorbed about half the CO_2 produced by combustion of fossil fuels and cement production leading to a reduction in the pH of surface seawater of 0.1 units, equivalent to a 30% increase in the concentration of hydrogen ions (Raven et al. 2005; Orr et al. 2009). It will take tens of thousands of years for ocean chemistry to return to a condition similar to that occurring at pre-industrial times of the 1790s. Reducing CO_2 emissions to the atmosphere appears to be the only practical way to minimize the risk of large-scale and long-term changes to the oceans (Raven et al. 2005). Surface ocean pH is projected to decrease by 0.4 pH units by the year 2100 relative to pre-industrial conditions (Orr et al. 2009). Marine organisms have not experienced this rate of acidification for millions of years (Orr et al. 2009). Current acidification rates differ from past events because current rates are human-induced and are occurring more rapidly. Records of ocean carbon chemistry over the past 25 years show clear trends of increasing ocean organic carbon and decreasing pH, closely paralleling increasing atmospheric CO_2 (Orr et al. 2009). Significant surface ocean acidification—with pH decreasing from 8.16 to 7.46—is predicted by the year 2300 (Montenegro et al. 2007).

Anthropogenic CO_2 in the world oceans is estimated at 112 (95–129) Pg C during the industrial era, accounting for about 29% of the total CO_2 emissions from the burning of fossil fuels, land-use changes, and cement production during the past 250 years (K. Lee et al. 2003). Ocean acidification describes the progressive global reduction in seawater pH because of the accelerating oceanic uptake of atmospheric carbon dioxide (Pelejero et al. 2005; Manzello et al. 2008; Turley et al. 2010). In the natural carbon cycle, the atmospheric concentration of carbon dioxide (CO_2) is balanced among the oceans, terrestrial biosphere, lithosphere, and the atmosphere. Anthropogenic activities—including fossil fuel combustion, land use changes, and cement production—have led to a new flux of CO_2 into the atmosphere. Some of the CO_2 has remained in the atmosphere (Table 1.1), some is taken up by terrestrial plants (Cramer et al. 2001), and some has been absorbed by the oceans (Raven and Falkowski 1999). When CO_2 dissolves it reacts with water to form ionic and non-ionic chemicals such as dissolved free carbon dioxide, carbonic acid (H_2CO_3), bicarbonate (HCO_3^-) and carbonate (CO_3^{2-}). According to Key et al. (2004), oceanic pH has declined from 8.179 in the pre-industrial period (1700s) to 8.104 in the 1990s. Uptake of CO_2 by the ocean since preindustrial times (circa late

Table 1.1. Atmospheric CO_2 between the years 1000 and 2006 (modified from Doney 2006).

Year	Concentration (parts per million)
1000	280
1200	285
1400	282
1600	275
1800	276
1900	286
1950	315
1960	320
2000	340
2006	375

1700s) has led to a current reduction of the pH of surface seawater of 0.1 units, equivalent to a 30% increase in the concentration of hydrogen ions (Raven et al. 2005). If global emissions of CO_2 from human activities continue to rise based on current trends, then the average pH of the oceans could fall by 0.5 units—equivalent to a three-fold increase in the concentration of hydrogen ions—by the year 2100 (Haugan and Drange 1996; Raven et al. 2005). Projected models by Orr et al. (2005) indicate that oceanic pH will be 7.949 in 2050 and 7.824 in 2100. The projected increase in deposition of atmospheric CO_2 into the oceans will result in a surface water pH decline, as measured logarithmically (Wolf-Gladrow et al. 1999), and this will dramatically impair calcification mechanisms of temperate and tropical corals and coralline macroalgae (Gattuso et al. 1998; Riebesell et al. 2000; Langdon 2002; Hoegh-Guldberg et al. 2007; USDC 2008).

It is acknowledged that production and dissolution of oceanic calcium carbonate ($CaCO_3$) has a major influence on the capacity of the oceans to absorb excess atmospheric carbon dioxide (Andersen and Malahoff 1977; Broecker et al. 1979; Fabry 1989). The major sources of oceanic $CaCO_3$ are the shells secreted by planktonic foraminiferans, coccolithophorids, and molluscan pteropods (Milliman 1974; Honjo 1977; Fabry 1989). Foraminiferans and coccolithophorids precipitate calcium carbonate in the form of calcite; however, pteropods produce shell of aragonite, a metastable form of $CaCO_3$ that is about 50% more soluble in seawater than calcite (Mucci 1983). When these organisms die, their carbonate skeletons either dissolve or accumulate in ocean sediments. Carbon dioxide that has diffused from the air into surface layers of the oceans eventually will settle to the bottom where it will be neutralized by carbonate particles, a process that takes 1,000 to 5,000 years (Broecker and Takahashi 1977; Broecker et al. 1979). Today's surface ocean is saturated with respect to calcium carbonate, but increasing atmospheric carbon dioxide concentrations are reducing ocean pH and carbonate ion concentrations, and thus the level of calcium carbonate saturation (Orr et al. 2005). If these trends continue, key marine organisms—including corals—will have difficulty maintaining their external calcium carbonate skeletons. Authors project that by 2050 the Southern Ocean surface waters will start to become undersaturated with respect to aragonite. By 2100, this undersaturation could extend throughout the entire Southern Ocean and into the subarctic Pacific Ocean. Conditions detrimental to high-latitude ecosystems could develop within decades, not centuries, as suggested previously (Orr et al. 2005).

Carbon dioxide-induced damage to coral reef ecosystems and the fisheries and recreation industries that depend on them could amount to economic losses of many billions of dollars annually (Raven et al. 2005; Bradshaw 2007; USDC 2008; Cooly et al. 2009; Feely et al. 2009). In the longer term, changes to the stability of coastal reefs may reduce the protection they offer to coasts. There may also be direct and indirect effects on commercially important species of fish and shellfish (Raven et al. 2005).

1.3 Terrestrial and Freshwater Acidification

Terrestrial and freshwater ecosystems are heavily impacted by acidic precipitation, including acid rain, snow, fog, and mist. Combustion of fossil fuels such as oil, coal, and gasoline are the acknowledged sources of the acidity. The waste sulfur, nitrogen, and carbon aerosols from these emissions combine with water to form sulfuric, nitric, and carbonic acids, which are subsequently transported by prevailing winds, sometimes thousands of kilometers from the initial source (USEPA 1980). Acid rain can damage property, fish, soil, and some crops. Potential effects of acid precipitation on non-oceanic ecosystems is dependent, in part, on amount and frequency of precipitation, soil-buffering capacity, type and abundance of vegetation, and degree of human activity such as automobiles, smokestacks, and other emitters of by-products of fossil fuel combustion (USEPA 1980). Precipitation in Europe and eastern North America has become acidic, a result of increases in acid-containing aerosols produced by fossil fuel combustion, metal smelting and industrial processes (Haines 1981). The increased use of tall smoke stacks and particle removers has increased long-range transport of acidic gases. In regions where acid-neutralizing capacity of soils and water is low, the pH of lakes and streams have decreased, concentrations of metals have increased, and freshwater biota of all taxa have been reduced in abundance. Freshwater fish populations of acidified ecosystems, as one example, have experienced death, species decline, reduced growth, skeletal deformities, and reproductive failure (Haines 1981). Acidic deposition is now reported in eleven U.S. western states, mostly from automobiles, utilities, and copper smelters in Mexico and Arizona (Sun 1985). Acidic deposition in these states occurs mainly in the form of dry microscopic particles, in addition to rain, fog, and snow. Western mountains are comparatively steep and have a thinner soil base than Eastern mountains. As a result, acid deposition may flow into lakes and streams without being neutralized as much as they would in the East where the terrain is more gently sloped and deeper soils permit runoff to percolate more deeply in the soils. Western mountains accumulate more snow and during a spring snow melt potentially deliver a burst of acidity to the surroundings. Western mountain ranges, including the Cascades, Rockies, and Sierra Nevadas are located downwind from sources of nitrogen oxides and sulfur dioxide emissions. Trees in west coast conifer forests may be especially susceptible to damage because they are exposed to high levels of acid deposition as well as ozone which together may weaken the health of trees (Sun 1985).

A proposal to effect a 30% reduction of sulfur dioxide emissions from 1980 emission levels by 1995 was endorsed by ten nations in 1984: Germany, Norway, Sweden, Denmark, Finland, Switzerland, Austria, Canada, France, and the Netherlands (Keating 1984). The United States, however, did not adopt the proposal at that time pending a review of options for dealing with acid precipitation (Keating 1984).

1.4 Literature Cited

Boyce, D.G., M.R. Lewis, and B. Worm. 2010. Global phytoplankton decline over the past century, *Nature*, 466, 591–596.

Bradshaw, K. 2007. Discovering the effects of CO_2 levels on marine life and global climate, *Sound Waves*, U.S. Geol. Surv. Monthly newsletter, Jan./Feb. 2007, 7 pp.

Broecker, W.S. and T. Takahashi. 1977. Neutralization of fossil fuel CO_2 by marine calcium carbonate. In N.R. Andersen and A. Malahoff (Eds.). The fate of fossil fuel CO_2 in the oceans. Plenum Press, New York, pp. 213–241.

Broecker, W.S., T. Takahashi, H.J. Simpson, and T.H. Peng. 1979. Fate of fossil fuel carbon dioxide and the global carbon budget. *Science*, 206, 409–418.

Chen, G.T. and F.J. Millero. 1979. Gradual increase of oceanic CO_2, *Nature*, 277, 205–206.

Cooley, S.R., H.L. Kite-Powell, and S.C. Doney. 2009. Ocean acidification's potential to alter global marine ecosystem services, *Oceanography*, 22, 172–181.

Cramer, W., A. Bondeau, F.I. Woodward, I.C. Prentice, R.A. Betts, V. Brovkin, P.M. Cox, V. Fisher, J. Foley, A.D. Friend, C. Kucharik, M.R. Lomas, N. Ramankutty, S. Stitch, B. Smith, A. White, and C. Young-Molling. 2001. Global response of terrestrial ecosystem structure and function to CO_2 and climate change results from six dynamic global vegetation models. *Global Change Biol.*, 7, 357–373.

Doney, S.C. 2006. The dangers of ocean acidification, *Sci. Amer.*, March 2006, 58–65.

Doney, S.C., V.J. Fabry, R.A. Feely, and J.A. Kleypas. 2009. Ocean acidification: the other CO_2 problem, *Ann. Rev. Mar. Sci.*, 1, 169–192.

Dupont, S., N. Dorey, and M. Thorndyke. 2010. What meta-analysis can tell us about vulnerability of marine biodiversity to ocean acidification? *Estuar. Coastal Shelf Sci.*, 89, 182–185.

Fabry, V.J. 1989. Aragonite production by pteropod molluscs in the subarctic Pacific, *Deep-Sea Res.*, 36, 1735–1751.

Fabry, V.J., B.A. Seibel, R.A. Feely, and J.C. Orr. 2008. Impacts of ocean acidification on marine fauna and ecosystem processes, *ICES J. Mar. Sci.*, 65, 414–432.

Feely, R.A., S.C. Doney, and S.R. Cooley. 2009. Ocean acidification. Present conditions and future changes in a high-CO_2 world, *Oceanography*, 22, 36–47.

Gattuso, J.P., M. Frankignoulle, I. Bourge, S. Romaine, and R.W. Buddemeier. 1998. Effect of calcium carbonate saturation of seawater on coral calcification, *Global Planet Change*, 18, 37–46.

Gillis, J. 2010. Reading Earth's future in glacial ice. Rising seas predicted as threat to coastal areas, *N.Y. Times* (Nov. 14, 2010) International Sect. 1, 14, 15.

Guinotte, J.M. and V.J. Fabry. 2008. Ocean acidification and its potential effects on marine ecosystems, *Ann. N.Y. Acad. Sci.*, 1134, 320–342.

Haines, T.A. 1981. Acidic precipitation and its consequences for aquatic ecosystems: a review, *Trans. Amer. Fish. Soc.*, 110, 669–707.

Haugan, P.M. and H. Drange. 1996. Effects of CO_2 on the ocean environment, *Energy Convers. Mgmt.*, 37, 1019–1022.

Hoegh-Guldberg, O., P.J. Mumby, A.J. Hooten, R.S. Steneck, P. Greenfield. E, Gomez, C.D. Harvell, P.F. Sale, A.J. Edwards, K. Caldeira, N. Knowlton, C.M. Eakin,

R. Iglesias-Prieto, N. Muthiga, R.H. Bradbury, A. Dubi, and M.E. Hatziolos. 2007. Coral reefs under rapid climate change and ocean acidification, *Science*, 318, 1737–1742.

Hofmann, G.E., J.P. Barry, P.J. Edmunds, R.D. Gates, D.A. Hutchins, T. Klinger, and M.A. Sewell. 2010. The effect of ocean acidification on calcifying organisms in marine ecosystems: an organism-to-ecosystem perspective, *Ann. Rev. Ecol. Evol. System*, 41, 127–147.

Honjo, S. 1977. Biogenic carbonate particles in the ocean; do they dissolve in the water column? In N.R. Andersen and A. Malahoff (Eds.). The fate of fossil fuel CO_2 in the oceans, Plenum Press, New York, pp 269–294.

Karl, T.R. and K.E. Trenberth. 2003. Modern global climate change, *Science*, 302, 1719–1723.

Keating, M. 1984. International Conference of Ministers on Acid Rain. Ottawa. Canada, March 20, 21, 1984. *The Globe and Mail*, Toronto, Canada. March 22, 1984.

Key, R.M., A. Kozyr, C.L. Sabine, K. Lee, R. Wanninkhof, J. Bullster, R.A. Feely, F. Millero, C. Mordy, and T.H. Peng. 2004. A global ocean carbon climatology: results from Global Data Analysis Project (GLODAP), *Global Biogeochemical Cycles*, 18, 2247–2295.

Langdon, C. 2002. Review of experimental evidence for effects of CO_2 on calcification of reef builders, *Proc. 9th Int. Coral Reef. Symp., Bali, Indonesia, Oct 23–27, 2000*, 2, 1091–1098.

Lee, K., S.D. Choi, G.H. Park, R. Wanninkhof, T.H. Peng, R.M. Key, C.L. Sabine, R.A. Feely, J.L. Bullister, F.J. Millero, and A. Kozyr. 2003. An updated anthropogenic CO_2 inventory in the Atlantic Ocean, *Global Biogeochem. Cycles*, 17(4), 1116, doi:10.1029/2003GB002067.

Manzello, D.P., J.A. Kleypas, D.A. Budd, C.M. Eakin, P.W. Glynn, and C. Langdon. 2008. Poorly cemented coral reefs of the eastern tropical Pacific: possible insights into reef development in a high CO_2 world *Proc. Natl. Acad. Sci. USA*, 105, 14050–10455.

Millero, F.J. 2007. The marine inorganic carbon cycle, *Chem. Rev.*, 107, 308–341.

Milliman, J.D. 1974. Marine carbonates. Springer-Verlag, New York, 375 pp.

Montenegro, A., V. Brovkin, M. Eby, D. Archer, and A.J. Weaver. 2007. Long term fate of anthropogenic carbon, *Geophys. Res. Lett.*, 34, L19709, doi:10.1029/2007GL030905.

Mucci, A. 1983. The solubility of calcite and aragonite in seawater at various salinities, temperatures and one atmosphere pressure, *Amer. J. Science*, 283, 780–799.

Orr, J.C., K. Caldeira, V. Fabry, J.P. Gattuso, P. Haugan, P. Lehodey, S. Pantoja, H.O. Portner, U. Riebesell, T. Trull, E. Urban, M. Hood, and W. Broadgate. 2009. Research priorities for understanding ocean acidification. Summary from the Second Symposium on the Ocean in a High-CO_2 World, *Oceanography*, 22, 182–189.

Orr, J.C., V.J. Fabry, O. Aumont, L. Bopp, S.C. Doney, R.A. Feely, A. Gnanadesikan, N. Gruber, A. Ishida, F. Joos, R.M. Key, K. Lindsay, E. Maier-Reimer, R. Matear, P. Monfray, A. Mouchet, R.G. Najjar, G.K. Plattner, K.R. Rodgers, C.L. Sabine, J.L. Sarmiento, R. Schlitzer, R.D. Slater, I.J. Totterdell, M.F. Weirig, Y. Yamanaka, and A. Yool. 2005. Anthropogenic ocean acidification over the twenty-first century and its impact on calcifying organisms, *Nature*, 437 (7059), 681–686.

Palacios, S.L. and R.C. Zimmerman. 2007. Response of eelgrass *Zostera marina* to CO_2 enrichment: possible impacts of climate change and potential for remediation of coastal habitats, *Mar. Ecol. Prog. Ser.*, 344, 1–13.

Pelejero, C., E. Calvo, M.T. McCulloch, J.F. Marshall, M.K. Ganan, J.M. Lough, and B.N. Opdyke. 2005. Preindustrial to modern interdecadal variability in coral reef pH, *Science*, 309, 2204–2207.

Portner, H.O., M. Langenbuch, and A Reipschlager. 2004. Biological impact of elevated ocean CO_2 concentrations: lessons from animal physiology and earth history, *J. Ocean.*, 60, 705–718.

Raven, J., K. Caldeira, H. Elderfield, O. Hoegh-Guldberg, P. Liss, U. Riebesell, J. Shepard, C. Turley, and A. Watson. 2005. Ocean acidification due to increasing atmospheric carbon dioxide, *Policy doc. 12/05, The Royal Society, 6–9 Carlton House Terrace, London SW15AG*, 57 pp.

Raven, J.A. and P.G. Falkowski. 1999. Oceanic sinks for atmospheric CO_2, *Plant Cell Environ.*, 22, 741–755.

Revelle, R. and H.E. Suess. 1957. Carbon dioxide exchange between atmosphere and ocean and the question of an increase of atmospheric CO_2 during the past decades, *Tellus*, 9, 18–27.

Riebesell, U., I. Zondervan, B. Rost, P.D. Tortell, R.E. Zeebe, and F.M.M. Morel. 2000. Reduced calcification of marine plankton in response to increased atmospheric CO_2, *Nature*, 407, 364–367.

Rijnsdorp, A.D., M.A. Peck, G.H. Engelhard, C. Mollmann, and J.K. Pinnegar. 2009. Resolving the effect of climate change on fish populations, *J. Mar. Sci.*, 66, 1570–1583.

Shellito, C.J., L.C. Sloan, and M. Huber. 2003. Climate model sensitivity to atmospheric CO_2 levels in he Early-Middle Paleogene, *Paleogeo, Paleoclim, Paleoecol.*, 193, 113–123.

Sun, M. 1985. Possible acid rain woes in the West, *Science*, 228, 34–35.

Turley, C. 2008. Impacts of changing ocean chemistry in a high-CO_2 world, *Mineralog. Mag.*, 72, 359–362.

Turley, C., M. Eby, A.J. Ridgwell, D.N. Schmidt, H.S. Findlay, C. Brownlee, U. Riebesell, V.J. Fabry, R.A. Feely, and J.P. Gattuso. 2010. The societal challenge of ocean acidification, *Mar. Pollut Bull*, 60, 787–792.

Tyrrell, T., J.G. Shepherd, and S. Castle. 2007. The long-term legacy of fossil fuels, *Tellus*, 59B, 664–672.

UNEP (United Nations Environment Programme). 2010. Environmental consequences of ocean acidification: a threat to food security, *Available from UNEP*, P.O. Box 39552, Nairobi 00100, Kenya, 12 pp.

United States Department of Commerce (USDC), National Oceanic and Atmospheric Administration (NOAA). 2008. State of the Science Fact Sheet. Ocean acidification, 2 pp.

United States Environmental Protection Agency (USEPA). 1980. *Acid Rain*. USEPA Report 600/9-79-036, 36 pp.

Weart, S.R. 2010. The idea of anthropogenic global climate change in the 20th century, *Wiley Interdisc. Rev, Climate Change*, 1, 67–81.

Widdicombe, S. and J.I. Spicer. 2008. Predicting the impact of ocean acidification on benthic diversity. What can animal physiology tell us?, *J. Exp. Mar. Biol. Ecol.*, 366, 187–197.

Wolf-Gladrow, D.A., U. Riebesell, S, Burkhardt, and J. Bijma. 1999. Direct effects of CO_2 concentration on growth and isotopic composition of marine plankton, *Tellus*, 51B, 461–476.

CHAPTER 2

Historical

2.1 General

The age of planet Earth is now estimated at 4.54 billion years (http://pubs.usgs.gov/gip/geotime/age.html), but modern humans, *Homo sapiens*, evolved in Africa 250,000 to 400,000 years ago having migrated from the continent only 50,000 to 100,000 years ago (en.wikipedia.org/wiki/Human_evolution). A review of atmospheric carbon dioxide and oceanic pH concentrations over geologic time scales from the present to 545 million years ago is briefly presented below. Over the past 200 years, the increase in the burning of forest fuels and changes in land use has increased atmospheric CO_2 concentrations from about 280 ppm to 385 ppm (Turley 2008). These are the highest rates of atmospheric CO_2 increase for at least 800,000 years, and possibly for the past 30 million years. Increasing anthropogenic CO_2 may result in increased acidity of the world's surface oceans, with rapid decreases in ocean pH predicted for this century, concomitantly with warming seas, creating multiple threats to the marine environment. The future addition of massive amounts of CO_2 to surface waters could have a profound impact on ocean chemistry and could have an equally profound impact on biogeochemical cycles and marine ecosystems (Turley 2008). Throughout Earth's history, the oceans have played a dominant role in the climate system through the storage and transport of heat and the exchange of water and climate-relevant gases with the atmosphere (Riebesell et al. 2009). The ocean's heat capacity is about 1,000 times larger than that of the atmosphere and its content of reactive carbon more than 60 times larger. Through a variety of interlinked and nonlinear processes, the ocean acts as a driver of climate variability on time scales ranging from seasonal to glacial-interglacial; however, the sign and magnitude of the ocean's carbon cycle to climate change is as yet unknown (Riebesell et al. 2009).

2.2 Present to 35,000 Years Ago

An annual net uptake flux of CO_2 by the global oceans for the year 1995 is computed to be 2.2 Pg C per year (1 Petagram = 10^{15} g) (Takahashi et al. 2002). A zone between 40° and 60° latitudes in both the northern and southern hemispheres is found to be a major sink for atmospheric CO_2. Small areas such as the northwestern Arabian Sea and eastern equatorial Pacific Ocean, where seasonal upwelling occurs, exhibit intense seasonal changes in CO_2 partial pressure (pCO_2) due to the biological drawdown of CO_2 (Takahashi et al. 2002). Mean distribution for the surface water pCO_2 over the global oceans has been constructed for a reference year 2000 based upon about 3 million measurements of surface water pCO_2 between 1970 and 2007 (Takahashi et al. 2009). Seasonal changes in the surface water pCO_2 and the sea-air pCO_2 over four climatic zones in the Atlantic Ocean, Pacific Ocean, Indian Ocean, and the Southern Ocean are presented. The equatorial Pacific is the major source for atmospheric CO_2 and the temperate oceans in both hemispheres are the major sink zones. The high-latitude North Atlantic Ocean, including the Nordic Sea and portions of the Arctic Sea are the most intense CO_2 sink areas with a mean of 2.5 tons of carbon per month/km^2 owing to a combination of low CO_2 in seawater and high gas exchange rates. The total annual ocean uptake flux including the anthropogenic CO_2 is estimated to be 2.0 Pg carbon (Takahashi et al. 2009).

A record of atmospheric CO_2 from 1006 to 1978 CE was produced by analyzing air enclosed in ice cores from Antarctica (Etheridge et al. 1996). Preindustrial CO_2 levels were in the range 275–284 ppm with lower levels occurring during 1550–1800 CE, probably as a result of the colder global climate. Major CO_2 increases occurred over the industrial period except during 1935–1945 when CO_2 mixing ratios stabilized or decreased slightly, probably as a result of natural variations of the carbon cycle on a decadal timescale (Etheridge et al. 1996). The 20th century increase in atmospheric CO_2, however, occurred more than an order of magnitude faster than any sustained change during the past 22,000 years (Joos and Spahni 2008). Simulated variability in atmospheric CO_2 over the past 1,000 years is dominated by terrestrial carbon flux variability, which in turn reflects regional changes in net primary production modulated by moisture stress (Doney et al. 2006). However, the oceans tend to damp by 20–25% slow variations in atmospheric CO_2 generated by the terrestrial biosphere (Doney et al. 2006).

During the last glacial period about 8,000 years ago, Earth's climate underwent frequent large and abrupt global changes (Broecker 1997). Atmospheric CO_2 concentrations were reduced by 80 ppm, but the reasons remain unknown (Heinze et al. 1991). Measurement of the $CaCO_3$ content and of foraminifera weights demonstrated that a prominent dissolution event occurred in the western equatorial Atlantic at the onset of the last glacial period (Broecker and Clark 2001). During the glacial and interglacial periods (8,000–35,000 years ago), estimated sea surface CO_2 concentrations were higher than those recorded in the Vostok ice core (40,000 years ago), suggesting that the Southern Ocean was a potential source of CO_2 to the atmosphere (Bentaleb et al. 1996). However, phytoplankton productivity areas in the Southern Ocean were, as observed today, not homogeneous, introducing uncertainty in determining the past status of the atmosphere-ocean equilibrium of the Southern Ocean (Bentaleb et al. 1996). Ocean biology contributed less than half the observed glacial-interglacial variations of 80 to 100 ppm in atmospheric CO_2 with physical processes responsible for most of the oceanic uptake of atmospheric CO_2 during glaciations (Kohfeld et al. 2005). Two periods, from 17,000–13,800 years ago and 12,300 to 11,200 years ago, are characterized by sustained rapid rates of atmospheric CO_2 increase of >12.0 ppm annually (Ridgwell et al. 2003). Because these periods are coincident with Southern Hemisphere deglaciation, authors aver that changes in the biogeochemical properties of the Southern Ocean surface are the most likely cause (Ridgwell et al. 2003).

2.3 From 35,000 to 450,000 Years Ago

Ice cores from East Antarctica (at the Vostok station) document climate and atmospheric history of the past 420,000 years and suggest that the Southern Ocean was important in regulating the long-term changes of atmospheric CO_2 (Petit et al. 1999). The succession of changes through each climate cycle and termination was similar, with atmospheric and climate properties oscillating between stable bounds. Atmospheric concentrations of carbon dioxide and methane correlate well with Antarctic air temperature throughout the record, thus supporting the idea that greenhouse gases have contributed significantly to the glacial-interglacial change. At present, atmospheric burdens of these two important greenhouse gases are the highest recorded during the past 420,000 years. As judged from the Vostok record, the long stable Holocene is a unique feature of climate during the past 420,000 years, with profound implications for evolution and the development of civilization (Petit et al. 1999). The recovery of deeper ice cores from the Vostok station in Antarctica extends the climate record to 740,000 years (Augustin et al. 2004). The transition from glacial to interglacial conditions about 430,000 years ago (Termination V) resembles the transition into the present interglacial period in terms of the magnitude of change in temperatures and greenhouse gases, albeit with significant differences. The interglacial stage following Termination V is exceptionally long (28,000 years) when compared to the 12,000 years recorded so far in the present interglacial period. Given the similarities between this earlier warm period and today, authors imply that without human intervention, a climate similar to the present one would extend well into the future (Augustin et al. 2004).

The deep-sea sediment oxygen isotopic composition is dominated by a 100,000 year cyclicity that is interpreted as the main ice-age rhythm (Shackleton 2000). At the 100,000 year period, atmospheric CO_2, Vostok air temperature, and deep-water temperature are in phase with orbital eccentricity. It is probably the response of the global carbon cycle that generates the eccentricity signal by causing changes in atmospheric carbon dioxide (Shackleton 2000).

2.4 From 450,000 to 2.1 Million Years Ago

A 2.1 million year record of sea surface partial pressure of CO_2, based on boron isotopes in planktonic foraminiferan shells, suggests that the atmospheric partial pressure of CO_2 was relatively stable before the mid-Pleistocene climate transition more than a million years ago (Honisch et al. 2009). Glacial CO_2 was higher before the transition, but interglacial CO_2 was similar to that of late Pleistocene interglacial cycles less than 450,000 years ago. These estimates are consistent with a close linkage between atmospheric CO_2 concentrations and global climate. However, there is no evidence that a long-term drawdown of atmospheric CO_2 during the interglacial period was the main cause of the climate transition (Honisch et al. 2009). The carbon dioxide content of the atmosphere affects the content of the surface ocean, which in turn affects seawater pH (Pearson and Palmer 1999). Based on boron isotopes in foraminifera, the pH of ancient seawater (7.91) from the tropical Pacific Ocean about 50 million years ago suggests that atmospheric pCO_2 (550 ppm) was similar to or slightly higher than current estimates (Pearson and Palmer 1999).

2.5 From 15 to 60 Million Years Ago

Based on ancient sedimentary rocks, abrupt climate changes occurred over the past 15 million years mainly as a result of polar melting. The ongoing buildup of greenhouse gases might trigger another of these oceanic reorganizations and thereby the associated large atmospheric changes (Broecker 1997). Based on stable carbon isotope values of di-unsaturated alkenones extracted from deep sea cores, a reconstructed record of CO_2 was prepared from the middle Eocene (45 million years ago) to the late Oligocene, 25 million years ago (Pagani et al. 2005). Atmospheric CO_2 ranged from 1,000 to 1,500 ppm in the late Eocene and declined thereafter reaching modern levels by late Oligocene. The fall in CO_2 likely allowed for a critical expansion of ice sheets on Antarctica (Pagani et al. 2005). Elevated Eocene atmospheric CO_2 and its subsequent decline is also documented by Lowenstein and Demicco (2006). Reconstruction of atmospheric CO_2 based on terrestrial leaf stomata from the middle Paleocene (55–60 million years ago) to the Eocene showed maximum values of 450 ppm CO_2 during the Paleocene, 350 ppm during the Eocene, and about 350 ppm during the Miocene, 16–18 million years ago (Royer et al. 2001). Overall, the Eocene ocean was probably more acidic than at present, as suggested by a shallower carbonate compensation depth at that time, implying that the

Eocene atmospheric CO_2 level was higher than it is today (Caldeira and Berner 1999).

The Paleocene-Eocene thermal maximum (PETM), which occurred about 55 million years ago, has been attributed to a sudden release of 1,500 gigatons of carbon dioxide and/or methane into the hydrosphere and atmosphere from massive flood basalt volcanism and the eruption of mid-ocean ridge basalt-like flows (Storey et al. 2007). The PETM is thought to have lasted 210,000 to 220,000 years. During the PETM, there was an abrupt decrease in the ^{13}C proportion of marine and sedimentary carbon, the sea surface temperature rose by 5°C in the tropics and more than 6°C in the Arctic, with some ocean acidification, and the extinction of 30% to 50% of the deep-sea benthic foraminiferal species (Storey et al. 2007). Atmospheric CO_2 levels in the early Eocene—50–56 million years ago—probably reached at least 2,240 ppm (Thrasher and Sloan 2009). This high concentration does not yield a regional climate—*viz.* western North America—that matches existing data, suggesting that high atmospheric CO_2 alone could not account for the early Eocene climate (Thrasher and Sloan 2009). A 60 million year record of atmospheric CO_2 was calculated based mainly on changes in the major ion composition of seawater, such as calcium and magnesium, and oscillations in the mineralogy of primary oceanic carbonate sediments (Demicco et al. 2003). Mean atmospheric CO_2 values ranged from 100 to 300 ppm with maximum values of 1,200 to 2,500 ppm between 60 and 40 million years ago. From 25 million years ago to the present, these values were 100 to 300 ppm, and were significantly lower than previous estimates made from seawater pH data where total dissolved inorganic carbon was assumed constant and in line with Tertiary atmospheric CO_2 concentrations of 65 million years ago (Demicco et al. 2003). Similar findings based on alkenone isotopes were reported by Zachos et al. (2001). Rapid global warming, and probably other stressors, at the end of the Paleocene was associated with mass extinctions (Thomas 2007).

2.6 From 89 To 545+ Million Years Ago

During the mid-Turonian (89 to 94 million years ago), calculated atmospheric CO_2 concentrations ranged from 1,450 to 2,690 ppm based on stable isotope composition of paleosol calcite (Sandler

2006). This high CO_2 level is similar to or somewhat higher than other estimates for the Cretaceous and in accord with calculated high Turonian temperatures of many studies (Sandler 2006).

Seawater chemistry has changed over the past 100 million years (Tyrrell and Zeebe 2004). Calcium concentrations have approximately halved, magnesium concentrations have doubled, surface ocean carbonate ion (CO_3^{2-}) has nearly quadrupled, and pH has increased from 7.5 to 8.1 during this interval. Causes of the changes are uncertain (Tyrrell and Zeebe 2004). Atmospheric CO_2 concentrations have varied significantly over the past 600 million years (Berner and Kothavala 2001). Based on a new model, atmospheric CO_2 concentrations were comparatively elevated during the early Paleozoic (Ordovician and Cambrian; about 500–600 million years ago = MYA), but experienced a large drop during the Devonian (350–400 MYA) and Carboniferous (285–360 MYA). Carbon dioxide concentrations were considerably higher during the early Mesozoic (200–245 MYA), with a gradual decrease to near current values within the past 50 million years (Berner and Kothavala 2001).

The Early Triassic (about 240 MYA) was characterized by anoxic deep ocean turnover, elevated CO_2, and shifts in ocean chemistry that reduced carbonate deposition (Stanley and Fautin 2001). The environment of the Middle Triassic , which was favorable to aragonite calcification, could have allowed soft-bodied ancestors of scleractinian corals to secrete skeletons. Scleractinian corals are important today and in the geologic past because of their prodigious ability to calcify. Modern corals exert important controls on global climate and the marine environment especially in the recycling of carbon (Stanley and Fautin 2001).

Over the past 545 million years, rates of diversification of marine fauna and levels of atmospheric CO_2 have been closely correlated (Cornette et al. 2002). The strength of the correlation (P<0.001) suggests that one or more environmental variables controlling CO_2 levels has a profound impact on evolution throughout the history of metazoan life. Further, the CO_2 levels are correlated with the dynamics of the origination and extinction of genera (Cornette et al. 2002).

2.7 Literature Cited

Augustin, L., C. Barbante, P.R.F. Barnes, J.M. Barnola, M. Bigler, E. Castellano, O. Cattani, J. Chappellaz, D. Dahl-Jensen, B. Delmonte, G. Dreyfus, G. Durand, S. Falourd, H. Fischer, J. Fluckiger, M.E. Hansson, P. Huybrechts, G. Jugie, S.J. Johnsen, J. Jouzel, P. Kaufmann, J. Kipfstuhl, F. Lambert, V.Y. Lipenkov, G.C. Littot, A. Longinelli, R. Lorrain, V. Maggi, V. Masson-Delmotte, H. Miller, R. Mulvaney, J. Oerlemans, H. Oerter, G. Orombelli, F. Parrenin, D.A. Peel, J.R. Petit, D. Raynaud, C. Ritz, U. Ruth, J. Schwander, U. Siegenthaler, R. Souchez, B. Stauffer, J.P. Steffensen, B. Stenni, T.F. Stocker, I.E.Tabacco, R. Udisti, R.S.W. van de Wal, M. Van den Broeke, J. Weiss, F. Wilhelms, J.G. Winther, E.W. Wolff, and M. Zucchelli. 2004. Eight glacial cycles from an Antarctic ice core, *Nature*, 429, 623–628.

Bentaleb, I., M. Fontugne, C. Descolas-Gros, C. Girardin, A. Mariotti, C. Pierre, C. Brunet, and A. Poisson. 1996. Organic carbon isotopic composition of phytoplankton and sea-surface pCO_2 reconstructions in the Southern Indian Ocean during the last 50,000 yr, *Org. Geochem.*, 24, 399–410.

Berner, R.A. and Z. Kothavala. 2001. Geocarb III: a revised model of atmospheric CO_2 over Phanerozoic time, *Amer. J. Sci.*, 301, 182–204.

Broecker, W.S. 1997. Thermohaline circulation, the Achilles heel of our climate system: will man-made CO_2 upset the current balance?, *Science*, 278, 1582–1588.

Broecker, W. and E. Clark. 2001. A dramatic dissolution event at the onset of the last glaciation, *Geochem. Geophys. Geosyst.*, 2, 1065, doi:10.1029/2001GC000185.

Caldeira, K. and R. Berner. 1999. Seawater pH and atmospheric carbon dioxide, *Science*, 286, 2043.

Cornette, J.L., B.S. Lieberman, and R.H. Goldstein. 2002. Documenting a significant relationship between macroevolutionary origination rates and Phanerozoic pCO_2 levels, *Proc. Natl. Acad. Sci. USA*, 99, 7832–7835.

Demicco, E.V., T.K. Lowenstein, and L.A. Hardie. 2003. Atmospheric pCO_2 since 60 Ma from records of seawater pH, calcium, and primary carbonate mineralogy, *Geology*, 31, 793–796.

Doney, S.C. 2006. The dangers of ocean acidification, *Sci. Amer.*, March 2006, 58–65.

Etheridge, D.M., L.P. Steele, R.L Langenfelds, R.J. Francey, J.M. Barnola, and V.I. Morgan. 1996. Natural and anthropogenic changes in atmospheric CO_2 over the last 1000 years from air in Antarctic ice and firn, *J. Geophys. Res.*, 101(D2), 4115–4128, doi:10,1029/95JD03410.

Gillis, J. 2010. Reading Earth's future in glacial ice. Rising seas predicted as threat to coastal areas, *N.Y. Times* (Nov. 14, 2010) International Sect., 1, 14, 15

Heinze, C., E. Maier-Reimer, and K. Winn. 1991. Glacial pCO_2 reduction by the world ocean: experiments with the Hamburg carbon cycle model, *Paleocean*, 6, 395–430.

Honisch, B., N.G. Hemming, D. Archer, M. Siddall, and J.F. McManus. 2009. Atmospheric carbon dioxide concentration across the mid-Pleistocene transition, *Science*, 344, 1551–1554.

Joos, F. and R. Spahni. 2008. Rates of change in natural and radiative forcing over the past 20,000 years, *Proc. Natl. Acad. Sci.* USA, 105, 1425–1430.

Kohfeld, K.E., C. Le Quere, S.P. Harrison, and R.F. Anderson. 2005. Role of marine biology in glacial-interglacial CO_2 cycles, *Science*, 308, 74–78.

Lowenstein, T.K. and R.V. Demicco. 2006. Elevated Eocene atmospheric CO_2 and its subsequent decline, *Science*, 313, 1928.

Pagani, M., J.C. Zachos, K.H. Freeman, B. Tipple, and S. Bohaty. 2005. Marked decline in atmospheric carbon dioxide concentrations during the Paleogene, *Science*, 309, 600–603.

Pearson, P.N. and M.R. Palmer. 1999. Middle Eocene seawater pH and atmospheric carbon dioxide concentrations, *Science*, 284, 1824–1826.

Petit, J.R., J. Jouzel, D. Raynaud, N.I. Barkov, J.M. Barnola, I. Basile, M. Benders, J. Chappellaz, M. Davis, G. Delaygue, M. Delmotte, V.M. Kotlyakov, M. Legrand, V.Y. Lipenkov, C. Lorius, L. Pepin, C. Ritz, E. Saltzman, and M. Stievenard. 1999. Climate and atmospheric history of the past 420,000 years from the Vostok ice core, Antarctica, *Nature*, 399, 429–436.

Ridgwell, A.J., A.J. Watson, M.A. Maslin, and J.O. Kaplan. 2003. Implications of coral reef buildup for the controls on atmospheric CO_2 since the Last Glacial Maximum, *Paleoceanography*, 18(4), 1083, doi:10.1029/2003PA000893.

Riebesell, U., A. Kortzinger, and A. Oschlies. 2009. Sensitivities of marine carbon fluxes to ocean change, *Proc. Natl. Acad. Sci. USA*, 106, 20602–20609.

Royer, D.L., S.W. Wing, D.J. Beerling, D.W. Jolley, P.L. Koch, L.J. Hickey, and R.A. Berner. 2001. Paleobotanical evidence for near present-day levels of atmospheric CO_2 during part pf the Tertiary, *Science*, 292, 2310–2313.

Sandler, A. 2006. Estimates of atmospheric CO_2 levels during th mid-Turonian derived from stable isotope composition of paleosol calcite from Israel, *Geol. Soc. Amer. Spec Pap.*, 416, 75–88.

Shackleton, N.J. 2000. The 100,000-year ice-age cycle identified and found to lag temperature, carbon dioxide, and orbital eccentricity, *Science*, 289, 1897–1902.

Stanley, G.D. Jr. and D.G. Fautin. 2001. Paleontology and evolution: the origins of modern corals, *Science*, 291, 1913–1914.

Storey, M., R.A. Duncan, and C.S. Swisher III. 2007. Paleocene-Eocene thermal maximum and the opening of the northeast Atlantic, *Science*, 316, 587–589.

Takahashi, T., S.C. Sutherland, C. Sweeney, A. Poisson, N. Metzl, B. Tilbrook, N. Bates, R. Wanninkhof, R.A. Feely, C. Sabine, J. Olafsson, and Y. Nojiri. 2002. Global sea-air CO_2 flux based on climatological surface ocean pCO_2 and seasonal biological and temperature effects, *Deep-Sea Res. II.* 49, 1601–1622.

Takahashi, T., S.C. Sutherland, R. Wanninkhof, C. Sweeney, R.A. Feely, D.W. Chipman, B. Hales, G. Friederich, F. Chavez, C. Sabine, A. Watson, D.C.E. Bakker, U. Schuster, N. Metzl, H. Yoshikawa-Inoue, M. Ishii, T. Midorikawa, Y. Nojiri, A. Kortzinger, T. Steinhoff, M. Hoppema, J. Olafsson, T.S. Arnarson, B. Tilbrook, T. Johannessen, A. Olsen, R. Bellerby, C.S. Wong, B. Delille, N.R. Bates, and H.J.W. de Baar. 2009. Climatological mean and decadal change in surface ocean pCO_2, and net sea-air CO_2 flux over the global oceans, *Deep-Sea Res. II*, 56, 554–577.

Thomas, E. 2007. Cenozoic mass extinctions in the deep sea: what perturbs the largest habitat on Earth?, *Geol. Soc. Amer. Spec. Paper*, 424, 1–23.

Thrasher, B.L. and L.C. Sloan. 2009. Carbon dioxide and the early Eocene climate of western North America, *Geology*, 37, 807–810.

Turley, C. 2008. Impacts of changing ocean chemistry in a high-CO_2 world, *Mineralog Mag.*, 72, 359–362.

Tyrrell, T. and R.E. Zeebe. 2004. History of carbonate ion concentration over the last 100 million years, *Geochim. Cosmochim. Acta*, 68, 3521–3530.

Zachos, J., M. Pagani, L. Sloan, E. Thomas, and K. Billups. 2001. Trends, rhythms, and aberrations in global climate 65 Ma to present, *Science*, 292, 686–693.

CHAPTER 3

Sources of Oceanic Acidification

3.1 General

At present, major sources of CO_2 include fossil fuel combustion and deforestation, while major sinks include the atmosphere and, from models, the oceans (Randerson et al. 1997; Millero 2007). Based on analysis of boron isotopes in foraminifera, it is alleged that oceanic pH has declined over the past 20 million years from about 9.0 to present day levels of 8.2 (Sanyal et al. 1995). During the latest ice age of the recent Pleistocene, which lasted about a million years, seawater pH during the post-glacial period in the most seriously affected areas dropped by 0.3 units to 7.94–7.96 (Sanyal et al. 1995); however, these results are considered preliminary. The global oceans contain about 50 times more inorganic carbon than does the atmosphere. During glacial periods, the oceans were sinks for atmospheric CO_2. During glacial-interglacial transitions the oceans were a source of CO_2 to the atmosphere, although mechanisms responsible for net CO_2 exchange between ocean and atmosphere remain unresolved (Raven and Falkowski 1999). The 20 ppm rise in atmospheric CO_2 content over the last 8,000 years was a consequence, in part, of the 500 Gt C increase in terrestrial biomass early in the present interglacial and an early Holocene demise of the supply of excess respiration CO_2 to sediment pore waters (Broecker et al. 2001).

Increasing CO_2 in seawater results in formation of carbonic acid (H_2CO_3), which causes acidification (Iglesias-Rodriguez et al. 2008).

Carbonic acid and water molecules form bicarbonate ions (HCO_3^-), reducing carbonic acid and the ocean's saturation rate of calcite, a form of calcium carbonate ($CaCO_3$) produced by coccolithophores. Elevated carbon dioxide also results in an increase in bicarbonate, the source of carbon for calcium in coccolithophores. The precipitation from seawater of $CaCO_3$, a basic substance, lowers pH. For this reason, and because a greater fraction of dissolved inorganic carbon (the sum of bicarbonate, carbonate, and aqueous carbon dioxide) is present as CO_2 at low pH, the formation of $CaCO_3$ in seawater stimulates an increase in the concentration of aqueous CO_2. Therefore, a decrease in marine calcification without an accompanying decrease in organic carbon export would lead to an increased drawdown of atmospheric CO_2 (Iglesias-Rodriguez et al. 2008). The organic carbon fluxes, which act as a sink for atmospheric carbon dioxide, are generally larger than the carbonate carbon fluxes working as a source but are comparable in the deep subtropical oceans (Tsunogai and Noriki 1991). Carbonate carbon fluxes are much smaller than those of the organic carbon fluxes, indicating that organisms producing carbonate particles exist rather evenly in the world ocean and that a substantial part of the carbonate produced is transported to the ocean bottom (Tsunogai and Noriki 1991).

Global inventory of dissolved inorganic carbon in the surface mixed layer from community production over an 8-month period during 1990 ranged from 6.7 to 8.0 Gt C. (1 Gt C = 1 x 10^{12} kg carbon = 1 billion metric tons); for the entire year these values were estimated to range between 9.1 and 10.8 Gt C (Lee 2001). Temperate and polar oceans from both hemispheres are the major sinks for atmospheric CO_2, whereas the equatorial oceans are the major sources of CO_2 (Takahashi et al. 1997). The Atlantic Ocean is the most important CO_2 sink, accounting for about 60% of the global ocean uptake, while the Pacific Ocean is neutral because of its equatorial source flux being balanced by the sink flux of the temperate oceans. The Indian Ocean and the Southern Ocean take up about 20% each (Takahashi et al. 1997). According to Fletcher et al. (2006), the greatest anthropogenic CO_2 uptake occurs in the Southern Ocean and the tropics. Uptake was greatest in the Southern Ocean and least in the Pacific Ocean, although considerable uncertainties remain in some regions, particularly the Southern Ocean. However, global anthropogenic CO_2 oceanic uptake—estimated at 2.2 Pg C per year—is strongly correlated with changes in the global CO_2 inventory (Fletcher et al. 2006). Annual mean sources and sinks of atmospheric CO_2 for the decade of the 1990s and the early 2000s is characterized by outgassing in the tropics, uptake in the mid-latitudes, and comparatively small fluxes in the high latitudes (Gruber et al. 2009). Estimates point toward a small sink in the Southern Ocean south of 44°S, a result of the near cancellation between substantial outgassing of natural CO_2 and a strong uptake of anthropogenic CO_2. Estimates are limited by failure to incorporate long-term changes in the ocean carbon cycle, such as the recent possible stalling in the expected growth of the Southern Ocean carbon sink (Gruber et al. 2009).

The potential exists for both positive and negative CO_2 feedbacks between the ocean and the atmosphere, including changes in both the physics (circulation, stratification) and biology (export production, calcification) of the ocean (Sabine et al. 2004) . These processes are not well understood. However, for the near future most of the known chemical feedbacks are expected to be positive. Thus, if the surface ocean CO_2 concentrations continue to increase in proportion with the atmospheric CO_2 increase, a doubling of atmospheric CO_2 from preindustrial levels is predicted to result in a 30% decrease in carbonate ion concentration and a 60% increase in hydrogen ion concentration. As the carbonate ion concentration decreases, the ocean's ability to absorb more CO_2 from the atmosphere is diminished (Sabine et al. 2004). By the end of the century, it is predicted that CO_2 levels in the oceans could be over 800 mg/L, the surface water dissolved inorganic carbon could probably increase by more than 12%, and the carbonate ion concentration would decrease by almost 60% (Feely et al. 2004). Such dramatic changes of the CO_2 system in open-ocean surface waters have not occurred for more than 20 million years, with biological effects imperfectly understood (Feely et al. 2004).

3.2 Anthropogenic

The growth rate of atmospheric CO_2 is increasing rapidly owing mainly to increasing combustion of fossil fuels worldwide, and evidence for a long-term increase (50 years) in the airborne fraction of CO_2 emissions—implying a decline in the efficiency of CO_2 sinks on land and oceans in absorbing anthropogenic emissions (Canadell et al. 2007). Since the year 2000, the contributions of these factors to the increase in atmospheric CO_2 growth rate have doubled. An increasing airborne fraction is consistent with results of climate-carbon models, but the magnitude of the observed signal appears larger than that estimated by models. All changes characterize a carbon cycle that is generating stronger-than-expected and sooner-than expected climate forcing (Canadell et al. 2007). Carbon dioxide emissions from fossil-fuel burning and industrial processes have been accelerating at a global scale, with their growth rate increasing from 1.1% annually for the years 1990–1999 to >3% for 2000–2004 (Raupach et al. 2007). At present, no region is decarbonizing its energy supply. The growth rate in emissions is strongest in rapidly-developing economies, particularly China. Together, the developing and least-developed economies (comprising 80% of the global population) accounted for 73% of global emissions growth in 2004 but only 41% of global emissions and only 23% of global cumulative emissions since the mid-eighteenth century (Raupach et al. 2007). Reactive sulfur and nitrogen discharged into the atmosphere from fossil fuel combustion and agriculture form dissociation products of strong acids (HNO_3 and H_2SO_4) and bases (NH_3) that can alter surface seawater alkalinity, pH, and inorganic carbon storage (Doney et al. 2007). On a global scale, however, the alterations in surface seawater chemistry from anthropogenic nitrogen and sulfur deposition are only a few percent of the acidification and dissolved inorganic carbon increases due to the oceanic uptake of anthropogenic CO_2. Impacts of reactive sulfur and nitrogen are probably more substantial in coastal waters, where ecosystem responses to ocean acidification are greater (Doney et al. 2007).

Over the past 250 years, the release of carbon dioxide from agriculture and industry has increased atmospheric CO_2 by about 100 ppm (parts per million), raising it to the highest level predicted for at least 650,000 years (Feely et al. 2008). Commercial airliners making trans-Atlantic flights also contribute to the CO_2 burden of the North Atlantic Ocean (Michaels 2010). A single one-way flight from North America to Europe burns more than 25 metric tons of fuel, producing approximately 75 tons of CO_2. At present, airlines operate more than 100,000 flights annually across the North Atlantic (Michaels 2010). Observations of CO_2 accumulation in the atmosphere and ocean show that they are approximately equal to the total amount emitted by combustion of fossil fuels since 1850, strongly suggesting that increased atmospheric CO_2 is wholly due to human activities and dominated by the burning of fossil fuels (Tans 2009). Although fossil fuel emissions enter the atmosphere, the CO_2 does not remain there. Nearly half the emissions to date have diffused across the sea-surface and entered the ocean. Because of the slow overturning of the ocean, most of the fossil fuel CO_2 now residing in the ocean is present in the surface layers (Tyrrell 2008). In locations where surface waters sink to depth, such as the high latitude North Atlantic, anthropogenic CO_2 penetrates the oceanic depths (Sabine et al. 2004). Current atmospheric concentrations are now approaching 380 ppm (Feely et al. 2004; Takahashi 2004; Doney 2006). Carbon dioxide is projected to approach 450 ppm by 2065, 650 ppm by 2100 (Palacios and Zimmerman 2007), and 2,000 ppm by the year 2200 (Millero 2007), leading to pronounced climate changes by the end of this century (Siegenthaler et al. 2005). Since the beginning of the industrial era (ca. 1790s), the oceans have absorbed 127 billion tons of carbon as CO_2 from the atmosphere, or about one-third of the anthropogenic carbon emitted (Sabine et al. 2004; Sabine and Feely 2007; Table 3.1). In the past few decades, only half the CO_2 released by human activity has remained in the atmosphere; of the remainder about 39% has been taken up by the oceans and 20% by the terrestrial atmosphere (Feely et al. 2004). On the time scale of several thousands of years, it is estimated that 90% of the anthropogenic CO_2 emissions will end up in the ocean (Archer et al. 1998); however, owing to the slow mixing time of the ocean this value at present remains at about 33 percent. During the

Table 3.1. Proposed anthropogenic CO_2 budget for the periods 1800–1994 and for 1980–1999 (modified from Sabine et al. 2004; Millero 2007).

CO_2 sources and sinks	1800–1994 (billions of tons)	1980–1999 (billions of tons)
Constrained sources and sinks		
Emissions from fossil fuel and cement production	244	117
Storage in the atmosphere	165	65
Uptake and storage in ocean	118	37
Inferred net terrestrial balance	39	15
Terrestrial balance		
Emissions for land-use change	140	24
Terrestrial biosphere sink	101	39

industrial revolution period (1790s) anthropogenic CO_2 accounted for 30 percent of oceanic CO_2 due, in part, to CO_2 solution in the surface ocean (Raven and Falkowski 1999). Models predict that future anthropogenic CO_2 inputs to the atmosphere will, in part, continue to be sequestered in the ocean. But changes in oceanic thermohaline circulation as a result of global climate change would greatly alter the predictions of carbon sequestration based on current circulation models (Raven and Falkowski 1999).

The uptake of anthropogenic CO_2 by the ocean was simulated using a three-dimensional model (Sarmiento et al. 1992). Atmospheric pCO_2 was prescribed for the period 1750 to 1990 based on ice cores from Antarctica and Hawaii. From 1980–1989 the average flux of CO_2 into the ocean was 1.9 billion metric tons of carbon annually, but this value may be low owing to limitations of the model. From 1980 to 1989, annual carbon input increased to 5.1 billion metric tons, and is comparable to the estimated fossil fuel CO_2 production of 5.4 Gt C annually—implying that other sources and sinks are approximately in balance (Sarmiento et al. 1992). The oceans' daily uptake of 22 million metric tons of carbon dioxide significantly affects its chemistry and biology, including a lowering of seawater pH by about 0.1 unit since the 1790s (Feely et al. 2008). Various models predict that atmospheric carbon dioxide concentrations could exceed 500 ppm by the year 2050 and 800 ppm by 2100. This increase would cause a decline in surface seawater pH of 0.4 units by the end of the century and a corresponding 50% decrease in carbonate ion concentration (Orr et al. 2005). These rapid pH changes are expected to negatively affect marine ecosystems (Raven et al. 2005). Sediment traps measure the rain of particulate matter through the water column and traps located at different depths at the same locations are used to estimate the dissolution rate of $CaCO_3$ particles between these two depths (Table 3.2). Dissolution rates in deep water are comparatively low and consistent with estimates of carbonate dissolution in the deep Pacific (Feely et al. 2004). Thus, of the total amount of $CaCO_3$ that is produced annually, no more than about 30% is buried in shallow and deep sediments. The rest is dissolved on the water column, at the sediment-seawater interface or in the upper portion of the sediment column. Results from Table 3.2 indicate that a very large fraction of this dissolution, 60% or more, occurs in the upper water column above 2,000 m (Feely et al. 2004).

Table 3.2. Sediment trap particulate dissolution fluxes in the Pacific Ocean[a] (modified from Feely et al. 2004).

Location	Trap depth range (m)	Dissolution rate (umol kg/ year)
Shallow sediment traps		
Northwestern Pacific	100–1000	0.12
Equatorial Pacific	105–320	0.67
Northwestern Pacific	500–1000	0.02
Northeastern Pacific	200–1000	0.10
Deep sediment traps		
Northwestern Pacific	2000–4000	0.003–0.006
Equatorial Pacific	3300–3600	0.005–0.014
2°59.8′ N, 135°1.0′ E	1592–3902	0.012
4°7.5′ N,136°16.6′ E	1769–4574	0.013

[a] The dissolution rates are derived from the differences in $CaCO_3$ sediment trap fluxes between the upper and lower sediment traps divided by the depth range between the traps

At the global scale, calcium carbonate plays a dual role in regulating carbon sequestration by the oceans (Feely et al. 2004). First, an increase in $CaCO_3$ dissolution in the upper ocean will result in a more uniform alkalinity profile with depth. A decrease in carbonate precipitation in the upper ocean would increase the capacity of the oceans to take up CO_2 from the atmosphere. A complete shutdown of surface ocean calcification would lower surface ocean CO_2. Second, a decrease in $CaCO_3$ production would affect the ratio of organic to inorganic carbon delivery to the deep sea. If the processes regulating the sinking of organic and inorganic carbon to deep sea sediments are uncoupled, then a decrease in calcium carbonate production would lead to increased dissolution of calcium carbonate sediments, which would raise ocean pH and its capacity to store CO_2. If the two processes are coupled, such as through the process of $CaCO_3$ ballasting of organic carbon, then reducing the carbonate production could result in shallower remineralization of organic carbon and a diminished role of sediments in buffering increases in atmospheric CO_2. Clearly more research is merited on the mechanistic controls of these seemingly coupled processes (Feely et al. 2004).

Riverine and estuarine CO_2 outgassing is significant globally (Zhai et al. 2005). In the case of the Pearl River estuary, China, various factors including season, organic pollution, and salinity all modify CO_2 production. Authors contend that more research is needed to evaluate the significance of CO_2 outgassing from these sources (Zhai et al. 2005). Coastal wetland soils frequently contain metal sulfides which oxidize in aerated conditions; sulfide oxidation produces sulfuric acid in drained coastal backswamp areas (Cook et al. 2000; Lin et al. 2004). Export of acidity from disturbed acidic sulphate soils will adversely impact inshore fisheries and breeding grounds with effects on fish and shellfish survival and histopathology (Cook et al. 2000). Rocky Mouth Creek, for example, located in eastern Australia, is the recipient of an artificial drain network that subsequently enters the nearby bay discharging acidic flows of pH 4.5 several months annually. This may be attributed to the hydrolysis of Fe^{3+} after the oxidation of Fe^{2+}; Fe^{2+} being generated by biological iron reduction, which consumes H^+ and thereby drives the conversion of retained acids to soluble acids. Installation of artificial drain works to improve soil conditions should be carefully monitored for acid outflow (Lin et al. 2004).

The impacts of increasing atmospheric CO_2 since the midst of the 18th century on seawater salinity and acidity are negligible, according to Loaiciga (2006). Assuming that the planetary mean surface temperature continues unabated, with the melting of terrestrial ice and permanent snow, the average seawater salinity would be lowered 0.61 ppt from its current 35 ppt and the increased atmospheric CO_2 would lower the seawater acidity by 0.09 pH units. A doubling of CO_2 to 760 ppm lowers the pH 0.19 units. Author concludes that over the time scales considered of hundreds of years, seawater salinity or acidity from observed or hypothesized rises in atmospheric CO_2 concentrations were minor (Loaiciga 2006). The findings of Loaiciga (2006) were dismissed in a comment signed by 24 respected authorities in the field of oceanic acidification (Caldeira et al. 2007). These authorities stated that Loaiciga made inappropriate assumptions, erroneous calculations, and mistaken conclusions. Specifically, the assumption of instantaneous chemical equilibration of the ocean with carbonate minerals, although this process is known to take 5,000 to 10,000 years and, contrary to what is implied by Loaiciga, many marine organism are indeed sensitive to a pH decrease of 0.2 units (Caldeira et al. 2007).

3.3 Biological

The role of photosynthetic organisms, invertebrates, and vertebrates in the production of CO_2 and other biogenic carbon species is briefly reviewed.

3.3.1 Photosynthetic Organisms

Marine phytoplankton are responsible for about half the global primary production and represent the basis of the marine food web (Rost et al. 2008). Phytoplankton drive various biogeochemical cycles, exporting massive amounts of carbon to deep waters and sediments, and strongly influence ocean-atmosphere gas exchanges. The increase in atmospheric CO_2 has now resulted in higher aquatic CO_2 concentrations and lower pH values. Responses of different phytoplankton species to the expected physico-chemical changes include changes in photosynthesis, calcification, and nitrogen fixation (Rost et al. 2008). Oceanic ecosystem processes are linked by net primary production in the surface layer, where inorganic carbon is fixed by photosynthetic processes (Behrenfeld et al. 2005). New satellite data show that chlorophyll:carbon ratios closely follow physiological dependencies on light, nutrients, and temperature. This fusion of emerging concepts from the phycological and remote sensing disciplines has the potential to change how we model and observe carbon cycling in the global oceans (Behrenfeld et al. 2005).

A model diagnosed the contribution of phytoplankton to the production and export of particulate organic carbon and $CaCO_3$ (Jin et al. 2006). The global export of $CaCO_3$ was estimated at 1.2 Pg C/year, with maxima at approximately 20°S, the equator, and at 40° N. The ratio of inorganic carbon to organic carbon was about 0.09, and uniform spatially. Larger phytoplankton dominated the export of particulate organic carbon, with diatoms contributing about 40% and coccolithophorids about 10%. Smaller phytoplankton dominated net primary production with a fraction of about 70%; diatoms contributed about 15% and coccolithophorids less than 2% (Jin et al. 2006). The export of biogenic carbon from the upper ocean is responsible for maintaining the vertical gradient of dissolved inorganic carbon, and thus indirectly for regulating the level of atmospheric CO_2 (Carlson et al. 1994). Measurements of dissolved organic carbon (DOC) near Bermuda demonstrated that DOC is an important component of the ocean carbon cycle. DOC accumulates in the early spring owing to increased primary productivity and is partially consumed in summer and autumn. The DOC that escapes remineralization is exported from the surface ocean the following winter and it is estimated that this export is at least equal to measured particle flux. Authors conclude that their observations are applicable to other parts of the world ocean which exhibit convective mixing and vernal restratification (Carlson et al. 1994). Based on empirical equations, phytoplankton particulate organic carbon production from environments in a high variance pigment statistical class is linearly related to chlorophyll-*a* within the near-surface layer (Iverson et al. 2000). These high variance environments are allegedly responsible for about 40% of global ocean annual phytoplankton carbon production and 70% of global ocean annual new and export production (Iverson et al. 2000). Surface waters with spectral signatures similar to that of *Emiliana huxleyi*, a coccolithophorid, annually cover an average of 1.4 X 10^6 km^2 (Buitenhuis et al. 1996). This species may influence global climate via 3 mechanisms: 1/by affecting the inorganic carbon system of seawater; 2/ by altering the heat exchange between seawater and the atmosphere, due to increased light scattering by detached coccoliths; and 3/by emissions of dimethylsulphide, which increased cloud albedo (Buitenhuis et al. 1996).

Carbon uptake by marine phytoplankton and its export as organic matter to the ocean interior lowers the partial pressure of carbon dioxide (pCO_2) in the upper ocean, and facilitates the drawdown of atmospheric CO_2 (Cermeno et al. 2008). Conversely, precipitation of $CaCO_3$ by marine plankton calcifiers, such as coccolithophorids, increases pCO_2 and promotes its outgassing. Over the past 100 million years or so these two carbon fluxes have been modulated by the relative abundance of diatoms and coccolithophorids. This balance is now threatened—according to a model formulated by the authors—that predicts a dramatic reduction in the nutrient supply to the euphotic zone within the next 100 years as a result of increased thermal stratification, causing a decrease in the sequestering of atmospheric CO_2 (Cermeno et al. 2008). Decreasing calcification and enhanced carbon overproduction by marine phytoplankton in response to rising atmospheric CO_2 levels and deliberate CO_2 sequestration in the ocean have the potential to increase the CO_2 storage capacity of the ocean; however, our ability to make reliable predictions of their future developments and to quantify their potential ecological and biogeochemical impacts are still unresolved (Riebesell 2004).

High variability of primary production in oligotrophic waters of Subtropical Gyres is reported for 1996–1997 (Maranon et al. 2003) and 1998 (Teira et al. 2003). These Gyres cover >60% of the total ocean surface and contribute >30% of the global marine carbon fixation. Variability of the North and South Atlantic Ocean Subtropical Gyres ranged from 18 to 362 mg C/m^2 daily while chlorophyll biomass only varied by a factor of 3 (Maranon et al. 2003). The rate of nutrient supply to the euphotic layer was the most relevant factor to account for the variability. It was concluded that the microbial community is the most important trophic pathway to account for carbon fixation variability in oligotrophic communities of the Atlantic Ocean (Maranon et al. 2003). Particulate organic carbon production in the eastern North Atlantic Subtropical Gyre in summer 1998 was, on average, 648 mg C/m^2 daily while dissolved organic carbon production rate was only 120 mg C/m^2 daily (Teira et al. 2003). The low dissolved organic carbon content is attributed to variability in trophic processes such as grazing and lysis (Teira et al. 2003). In the Atlantic Ocean during 2003–2004 surface rates of carbon fixation by phytoplankton ranged from <0.2 mmol C/m^3 daily in the subtropical gyres to 0.2–0.5 mmol C/m^3 daily in the tropical equatorial Atlantic (Poulton et al. 2006). Picoplankton (< 2 um) represented the dominant fraction in terms of both carbon fixation (50–70%) and chlorophyll-*a* (80–90%), but nannoplankton (>2 um) contributions to total carbon fixation (30–50%) were higher than total chlorophyll-*a* (10–20%). Differences between pico- and nannoplankton contributions to chlorophyll-*a* and carbon fixation appear to follow different patterns in surface waters (Poulton et al. 2006).

Phytoplankton biomass and rate of carbon uptake was determined in the northern North Sea in June 1999 within a bloom of the coccolithophore *Emiliana huxleyi* (Rees et al. 2002). Carbon fixation into particulate organic material ranged between 0.23 and 4.02 $umol$ C/L daily. Calcification, the production of inorganic carbon by coccolithophores, reached a maximum of 11.54 mmol C/ m^2 daily when *Emiliana huxleyi* cell numbers reached 1,500 cells/ mL (Rees et al. 2002). An extensive bloom (250,000 km^2) of the coccolithophore alga *Emiliana huxleyi* developed in the northeast Atlantic Ocean in June 1991 with high particulate inorganic values in excess of 300 mg/m^3 and calcification rates up to 2.5 mg C/m^3 per hour (Fernandez et al. 1993). Precipitation of calcium carbonate by phytoplankton in the photic oceanic layer is an important process regulating the carbon cycling and the exchange of CO_2 at the ocean-atmosphere interface (Sciandra et al. 2003). Each day, more than a hundred million tons of carbon in the form of CO_2 are fixed into organic material by upper ocean phytoplankton, and each day a similar amount of organic carbon is transferred into marine ecosystems by sinking and grazing (Behrenfeld et al. 2006). During glacial periods of increasing oceanic CO_2, phytoplankton primary productivity increased, leading to enhanced sedimentation and particulate organic carbon into the ocean interior (Raven and Falkowski 1999). Heterotrophic oxidation of the newly-formed organic carbon, forming weak acids, hydrolyze sediment $CaCO_3$, increasing oceanic alkalinity which, in turn, promotes the drawdown of atmospheric CO_2; this is consistent with stable carbon isotope patterns derived from air trapped in ice cores (Raven and Falkowski 1999). Carbon dioxide that is used for photosynthesis in *Emiliana huxleyi* comes from two sources. The CO_2 in seawater supports a minor photosynthetic rate; however, the HCO_3^- in seawater is the major substrate for photosynthesis by intracellular production of CO_2. At present, the HCO_3^- is the only substrate for calcification (Buitenhuis et al. 1999). In that study, the authors showed that calcification allows *E. huxleyi* to efficiently use both HCO_3^- and CO_2 in photosynthesis and this is reflected in growth rate (Buitenhuis et al. 1999). However, plankton manipulation experiments show a wide range of sensitivities of biogenic calcification to simulated anthropogenic acidification of the ocean, with the primary sentinel organism, *Emiliana huxleyi*, apparently not representative of calcification (Ridgwell et al. 2007). The implications of this observational uncertainty affects models of calcification response to ocean acidification, all of which are dominated by the assumption as to which species of calcifier contribute most to carbonate production in the open ocean (Ridgwell et al. 2007).

Most coral reefs are considered by some investigators to be sources of CO_2 to the atmosphere, contributing about 0.4 to 1.4% of the current anthropogenic CO_2 production due to fossil fuel combustion (Ware et al. 1991; Gattuso et al. 1993, 1999a; Chisholm and Barnes 1998). Because the precipitation of $CaCO_3$ results in the sequestering of carbon, some investigators believe that coral reefs function as sinks of global atmospheric CO_2. However, the precipitation of $CaCO_3$ is accompanied by a shift of pH that results in the release of CO_2 (Ware et al. 1991). A different viewpoint is presented by Gattuso et al. (1997). They studied community metabolism and air-sea CO_2 fluxes in July 1992 at a fringing reef in Moorea, French Polynesia. The benthic community was dominated by macroalgae, including several species of coralline algae. This site was a sink for atmospheric CO_2, and in this respect was typical of algal-dominated reefs (Gattuso et al. 1997). Measurement of air-sea CO_2 fluxes in open water near the fringing reef demonstrated that small hydrodynamic changes can lead to misleading conclusions. In this case, net CO_2 evasion to the atmosphere was measured on the fringing reef (Gattuso et al. 1993) due to changes in the current pattern that drove water from the barrier reef (a CO_2 source) to the study site (Gattuso et al. 1997). Primary productivity of 4 species of crustose algae was measured on samples from the windward coral reef at Lizard Island, northern Great Barrier Reef, Australia (Chisholm 2003). Net carbon fixation varied from 0.2 to 1.3 g/m² daily with estimated contributions to reef organic production of 0.9 to 5.0 g C m² daily over the depth interval 0 to 18 m. These data suggest that crustose coralline algae make a larger contribution to organic production on coral reefs than previously believed (Chisholm 2003). A prediction of the effect of photosynthetic and calcifying systems on air-sea CO_2 exchange at all levels from organism to ecosystem is based on air-sea CO_2 flux in a coral reef system (Gattuso et al. 1995). Input data for this model are: gross primary production (Pg), respiration (R), net calcification (G), and the ratio of CO_2 released to $CaCO_3$ precipitated (ψ); the output is the amount of dissolved inorganic carbon (F_{CO2}) which needs to be exchanged with the atmosphere to balance biologically- mediated changes in the concentration dissolved inorganic carbon in an open sea water system:

$$F_{CO2} = -Pg + R + \psi G.$$

A coral reef comprised of calcareous and non-calcareous organisms acts as a sink for atmospheric CO_2 when net production is high and $CaCO_3$ production is low. These characteristics are not typical of developing reef systems which exhibit a nearly balanced organic carbon metabolism (Pg/R = 1). This prediction is confirmed by limited field data (Gattuso et al. 1995).

3.3.2 Invertebrates

The roles of protists. coral reef communities, molluscs, crustaceans, echinoderms, and tunicates in the oceanic carbon cycle are briefly summarized.

3.3.2.1 Protists

Microorganisms, especially bacteria, significantly affect the oceanic carbon cycle constituting a major biological force in the ocean; their role, together with grazing food chains and sinking, require clarification (Azam 1998; Maranon et al. 2003). The potential importance of the viral lysis of phytoplankton for nutrient and carbon cycling is acknowledged (Gobler et al. 1997). Viral lysis of the chrysophyte, *Aureococcus anophagefferens*, with subsequent release and bioavailability of carbon, nitrogen, phosphorus, selenium, and iron is documented. Viral lysis of the chrysophyte released 50% more C and Se than uninfected controls to the dissolved phase, while N, P, and Fe remained in the particulate phase. Dissolved nutrients released by viral lysis were accumulated by marine bacteria and diatoms, and virally regenerated N and P relieved diatom nutrient limitation. Photochemical degradation of cell lysis reduced total dissolved C by 15%; photochemistry decreased the bioavailability of C to bacteria by a factor of three. It was concluded that field occurrences of viral lysing of chrysophytes may significantly affect water column chemistry, species composition, and succession within marine plankton communities (Gobler et al. 1997).

3.3.2.2 Coral Reef Communities

Community metabolism was investigated on two reef flats: one at Moorea in French Polynesia during austral winter, and another at

Yonge Reef on the Great Barrier Reef of Australia during austral summer (Gattuso et al. 1996). Community gross primary production and respiration were within the range previously reported for reef flats but community net calcification was higher than expected. The molar ratio of organic to inorganic carbon uptake was 6:1 for both sites. The daily air-sea flux was positive at all times indicating that the reef flats at Moorea and Yonge Reef released CO_2 to the atmosphere, although this decreased with increasing daily irradiance (Gattuso et al. 1996). Further study by Gattuso and his coworkers (Gattuso et al. 1999) show that coral reefs are important benthic, photosynthetic, calcifying ecosystems. They display a great diversity of photosynthetic $CaCO_3$-depositing organisms (calcareous algae) or harbor photosynthetic symbionts (reef-building corals, foraminiferans, molluscs). Large fluxes of carbon and calcium carbonate occur at the cell and community levels on reefs, The coral host is essential in supplying carbon for the photosynthesis of algal symbionts via carbon-concentration mechanisms described in free living algae; metabolic CO_2 seems to be a significant source, although the process remains murky. The rate of calcification decreases with increasing CO_2 and decreasing calcium carbonate saturation state. The calculated decrease in $CaCO_3$ production using a model constructed by the Intergovernmental Panel on Climate Change is 10% between 1880 and 1990, and 22% (9–30%) from 1990 to 2100. Inadequate understanding of the mechanism of calcification and its interaction with photosynthesis limits the ability for predictions of future calcification rates (Gattuso et al. 1999). Because the upper ocean is supersaturated with respect to all phases of $CaCO_3$ (Table 3.2), carbonate chemistry was not previously considered a limiting factor in biogenic calcification. However, the calcification rate of all calcifying organisms to date decreased in response to a decreased $CaCO_3$ saturation rate even when carbonate saturation level was >1 (Feely et al. 2004). This response holds across multiple taxa from single-celled protists to reef-building corals and across all $CaCO_3$ mineral phases (Feely et al. 2004).

Air-sea CO_2 exchanges and the partial pressure of CO_2 were measured in surface water overlying two coral reefs: one with low coral diversity and cover, the other with high coral diversity and cover (Frankignoulle et al. 1996). Both sites were net sources of CO_2 to the atmosphere as a result of the effect of calcification on the inorganic carbon system. At both sites the major exchange of CO_2

from sea to air occurs as seawater returns to chemical equilibrium after it has crossed and left the reef. About 80% of the change in inorganic carbon was related to photosynthesis and respiration, indicating that the calcification rate was proportional to the net organic production during the day and to the respiration rate at night (Frankignoulle et al. 1996). Complex physical and biological processes control the exchange of CO_2 between the ocean and atmosphere (Bates 2002). In coral reef ecosystems, most of the biological processes such as $CaCO_3$ formation and organic carbon production can either lead to CO_2 being retained in the oceanic environment or returned to the atmosphere through gas exchange, although the influence of seasonality and air-sea CO_2 fluxes is uncertain. A Bermuda coral reef ecosystem, for example, acts as source of CO_2 to seawater but varies seasonally in response to changes in the reef community reflecting changes in the net balance between calcification and organic carbon production. Whether the Bermuda coral reef system acts as an oceanic sink or source of CO_2 to the atmosphere depends on this seasonal variation and also the pre-existing air-sea CO_2 disequilibrium of open ocean waters surrounding the reef system (Bates 2002). Sea urchins (*Echinometra mathaei, Diadema savignyi, Echinothrix* spp.) and parrot fish (*Chlorurus sordidus*) were the most important grazers of reefs on Reunion Island in the Indian Ocean and Moorea Island in French Polynesia (Peyrot-Clausade et al. 2000). The total erosive activity of grazers for each location was about 8 kg $CaCO_3/m^2$ annually. Urchins were most important in bioerosion of the barrier reef flats, and fish on the fringing reef (Peyrot-Clausade et al. 2000).

3.3.2.3 Molluscs

The pteropod *Limacina helicina* produces 71,000 to 362,000 fecal pellets/m^2 in Terra Nova Bay, Ross Sea, Antarctica, with maximum values in March-April every year (Manno et al. 2010). The fecal pellet flux of this organism alone contributes 19% of the total particle organic carbon flux in Terra Nova Bay. The carbon pump may be modified if the pteropod population declines as a consequence of the predicted acidification in polar and subpolar waters (Manno et al. 2010). Fluxes of biogenic carbonates moving out of the euphotic zone and into deeper undersaturated waters of the North Pacific point to a major involvement in the oceanic carbonate system by aragonite pteropod molluscs; in fact, pteropod fluxes through the base of the euphotic zone are almost large enough to balance the alkalinity budget for the Pacific Ocean (Betzer et al. 1984).

3.3.2.4 Crustaceans

In subantarctic water, the Subtropical Front, and waters immediately to the north, the copepod *Neocalanus tonsus* makes a contribution to downwards carbon flux of 1.7–9.3 g C/m^2 annually (Bradford-Grieve et al. 2001). This flux is an order of magnitude greater than that estimated (0.27 g C/m^2) for vertical migration of large copepods in the North Atlantic. Over the range of *N. calanus*, 0.17 Gt C/year are estimated to be lost annually to the ocean interior and accounts for 1.4% of primary production in subantarctic waters. Different flux rates in different parts of its range may be due to nutrient status of these oceans, differences in the rate of development of grazer populations in the spring, and differences in life history characteristic of large copepods (Bradford-Grieve et al. 2001).

3.3.2.5 Echinoderms

Production of calcium carbonate was calculated from a large population of the brittle star, *Ophiothrix fragilis*, in Dover Strait, English Channel (Migne et al. 1998). Annual production was estimated at 682 g of $CaCO_3/m^2$, resulting in the release of 4.8 mol CO_2/m^2 yearly, reinforcing the suggestion that this coastal system is a source of CO_2 to the atmosphere (Migne et al. 2008).

The global contribution of echinoderms to the marine carbon cycle has been seriously underestimated (Lebrato et al. 2010). For example, echinoderm $CaCO_3$ production per unit area is estimated at 27.01 g $CaCO_3$ per m^2 (3.24 g C per m^2 annually as inorganic carbon) on a global scale for all areas, with a standing stock of 63.34 g $CaCO_3/m^2$ (7.60 g C/m^2 as inorganic carbon and 7.97 g C/m^2 as organic carbon). More than 80% of the global $CaCO_3$ production from echinoderms occurs between the surface and 800 m, with the highest contribution attributed to the shelf and upper slope (Lebrato et al. 2010).

During feeding on dead corals, echinoids removed a large proportion of algae and calcium carbonate and contributed significantly to the turnover of organic and inorganic carbon on coral reefs (Carreiro-Silva and McClanahan 2001). Rates of erosion of dead coral substratum, referred to as bioerosion, by four species of sea urchins on three Kenyan reefs were measured. One reef had been protected for more than 25 years against all forms of fishing as well as coral and shell collection; a second had been protected for the past 8 years; and the last was an unprotected reef with heavy fishing and some coral collection. Highest sea urchin densities (6.2/ m^2) were recorded in the unprotected reef with annual bioerosion estimated at 1,180.0 g $CaCO_3/m^2$; sea urchin densities (0.06/m^2) at protected reefs were 20 times lower than the unprotected reef with annual bioerosion estimated at 50.3 g $CaCO_3/m^2$; the newly-protected reef ,with an intermediate number of sea urchins (1.2/ m^2) showed an intermediate bioerosion rate of 711.0 g $CaCO_3/$ m^2 annually. Echinoids are important in the carbon cycle and reef development, but fishing can affect these processes (Carreiro-Silva and McClanahan 2001).

3.3.2.6 Tunicates

Planktonic tunicates play an important role in carbon transport to the sea floor (Robison et al. 2005). In a ten-year study of the water column off Monterey Bay, California, discarded mucus feeding structures of giant larvacean appendicularia, including *Bathochordaeus* sp., carry a substantial portion of the upper ocean's productivity to the deep seabed. These abundant, rapidly sinking, carbon-rich vectors should be considered in oceanic carbon budgets (Robison et al. 2005).

3.3.3 Vertebrates

Oceanic production of calcium carbonate is usually attributed to marine coccolithophores and foraminiferans. However, marine teleosts produce precipitated carbonates within their intestines and excrete these at high rates (Wilson et al. 2009). Based on global fish biomass, this suggests that marine fish contribute 3 to 15% of total oceanic carbonate production. Fish carbonates have a higher magnesium content and solubility than traditional sources, yielding faster dissolution with depth. Authors predict that fish carbonate production may rise with future environmental changes in CO_2 and thus become an increasingly important component of the inorganic carbon cycle (Wilson et al. 2009). In the gulf toadfish, *Opsanus beta*, and probably other marine teleosts, the ionic byproducts of osmoregulation contribute to the formation of a carbonate mineral in the intestine which, on excretion, is a substantial source of marine carbonate sediments (Walsh et al. 1991).

3.4 Physical

Volcanos produce about 200 million metric tons of CO_2 worldwide every year (www.guardian.co uk). The eruption of the Eyjafjallajokull volcano in Iceland during spring 2010 emitted about 150,000 metric tons daily as well as dense clouds of ash, effectively grounding thousands of trans-Atlantic flights. Since the European aviation industry emits about 344,000 tons of CO_2 daily, the eruption at Eyjafjallajokull resulted in a net reduction of almost 200,000 tons of CO_2 discharged into the environment daily (www.guardian. co uk).Through ocean ridge volcanism, deep-sea hydrothermal vents emit low pH fluids (near pH 3) enriched in CO_2 relative to seawater (Shitashima 1997). Estimates of CO_2 fluxes to the ocean through ocean ridges was calculated at 0.7 to 15×10^{12} mol C/year; these values were more than 1,000 times smaller than the annual CO_2 fluxes via terrestrial and marine respiration and input from this source was minimal on a short-term time scale. At longer time scales of 10^6–10^7 years, however, the flux of CO_2 from deep-sea hydrothermal systems may be significant (Shitashima 1997). Hydrothermal vents on oceanic sea mounts are a significant source of CO_2 in various locations (Vetter and Smith 2005). Hydrothermal vents on the Loihi seamount near Hawaii, for example, each emit up to 100,000 metric tons of CO_2 annually from water depths of 1,200–1,300 m at a mean pH of 6.3 (Vetter and Smith 2005).

Much of the coastal lowlands of eastern Australia are underlain by sulfidic sediments and large areas have been drained for agriculture (White et al. 1997) Drained sulfidic sediments oxidize and produce highly acidic waters (pH <4.0) with potential adverse effects on estuarine ecosystems. The rate of production from drained flood plains can be as high as 300 kg H_2SO_4/ha/year and hundreds of tons of H_2SO_4 can be discharged in a single flood on the plain. Generation and export of acidity is controlled by the water balance of the plain, the characteristics of the drainage system, and the distribution of sulfides. Evapotranspiration by native plants and crops play a dominant role in sediment oxidation during dry periods. In wet periods, upland discharges to flood plains dominate the water balance. Drain spacing and drain depth are critical factors in the export of acidity into coastal streams, and forms the basis of a mitigation strategy. This, and reflooding of unproductive acidified lowlands offers promise for rehabilitation of wetlands (White et al. 1997). Trends in the acid concentration of river water delivered to the ocean are spatially variable over continental scales (Salisbury et al. 2008). In recent decades, there was a significant shift in stream pH towards the acidic range throughout eastern Asia. However, acidity levels in precipitation and riverine discharges in North America have decreased due to reduced sulfate emissions. Anthropogenic patterns of land use and fertilization will also alter mineral dissolution and fluxes of carbonates germane to coastal acidification issues. The average annual discharge of fresh water from the six largest Eurasian rivers to the Arctic Ocean increased by 7% from 1936 to 1999; during that period, discharge was correlated with changes in global mean surface air temperature, dam operation, thawing of permafrost, as well as climate change (Salisbury et al. 2008).

During a survey of the end phase of an *Emiliana huxleyi* bloom in the northern part of the North Sea, total inorganic carbon, CO_2, $CaCO_3$, and particulate organic carbon were measured (Buitenhuis et al. 1996). Production of $CaCO_3$ resulted in an immediate increase of CO_2, but over time led to a long-term decrease in CO_2. This decrease is due to an enhanced sedimentation of both organic and inorganic carbon in fecal pellets containing heavy calcite. The enhanced sedimentation is reflected in the vertical gradient of total inorganic carbon between the surface mixed layer and the aphotic zone, which increased for the particulate organic carbon-rich zone to the $CaCO_3$ maximum. The overall effect of production, air-sea exchange, mineralization and sedimentation was a decrease of CO_2 due to a net transport of carbon below the pycnocline. Authors indicate that this *E. huxleyi* bloom alone represented an atmospheric carbon sink of 1.3 mol/m^2 (Buitenhuis et al. 1996). Fecal pellets of copepods and other zooplankton facilitate the removal of phytoplankton carbon from the water column over the continental shelf of the southern Middle Atlantic Bight during the spring (Lane et al. 1994). Diel migrant plankton and nekton obtain organic carbon by feeding at night above the main pycnocline of oceans and respire part of it by day in the interior of the ocean below the pycnocline (Longhurst et al. 1990). This flux of respiratory carbon ranged from 20 to 430 mg C/m² daily or 13 to 58% of computed particulate sinking flux across the pycnocline. If this flux occurs consistently between 50° N and 50° S, it will add about 5% to 30% to current estimates of global sinking flux of organic carbon across the pycnocline (Longhurst et al. 1990).

Future climate change will reduce the efficiency of the earth system to absorb the anthropogenic carbon perturbation (Friedlingstein et al. 2006). A larger fraction of anthropogenic CO_2 will stay airborne if climate change is considered. By the end of the 21st century, this additional CO_2 varied between 20 and 200 ppm for the two extreme models with the majority of the models ranging from 50 to 100 ppm. The higher CO_2 levels led to an additional climate warming of 0.1 to 1.5°C (Friedlingstein et al. 2006). Ocean warming or circulation alterations induced by climate change has the potential to slowdown the rate of acidification of ocean waters by decreasing the amount of CO_2 uptake by the ocean (McNeil and Matear 2006). However, climate change effects are insignificant with decreases in pH due to ocean warming and are balanced by the reduced dissolved inorganic carbon concentration of the upper ocean caused by the lower solubility of CO_2. The only way to mitigate the potential biological consequences of future ocean acidification is to significantly reduce fossil-fuel emissions of CO_2 to the atmosphere (McNeil and Matear 2006).

Total alkalinity of Atlantic, Pacific, and Indian Ocean surface water is modified by salinity, temperature, latitude, and dissolution of $CaCO_3$ (Millero et al. 1998). The distribution of surface alkalinity in the open ocean is mainly controlled by the factors that govern salinity. Salinity-normalized alkalinity in subtropical gyres between 30° S and 30° N is little changed except in upwelling areas. Salinity-normalized alkalinity increases with high latitudes (>30°) and is inversely proportional to sea surface temperatures; these increases are attributed to the upward transport of deep waters with higher dissolved $CaCO_3$ concentrations and the photosynthetic consumption of nutrients (Millero et al. 1998).

The oceans represent a significant sink for carbon dioxide, but there is great variability in the strength of this sink (Dore et al. 2003). In a 13-year study of oceanic CO_2 measurements in the subtropical North Pacific Ocean at a station near Hawaii, there was a significant decrease in the strength of the carbon dioxide sink between 1989 and 2001. Much of this reduction in sink strength is attributed to an increase in the partial pressure of surface oceanic carbon dioxide caused by excess evaporation and the concentration of solutes in the water mass. It was concluded that carbon dioxide uptake by ocean waters is strongly influenced by changes in regional precipitation and evaporation patterns caused by climate variability (Dore et al. 2003). Ocean waters are stratified according to their density (Takahashi 2004). In subsurface regimes, waters flow from polar regions toward lower latitudes along constant density horizons with little mixing between different densities. A parcel of water found at depths was previously located near the sea surface where it acquired CO_2 from the overlying atmosphere, from the oxidation of biogenic debris and dissolved organic compounds, and from the dissolution of skeletal $CaCO_3$ falling through the water column (Takahashi 2004).

3.5 Literature Cited

Archer, D.E., H. Kheshgi, and E. Maier-Reimer. 1998. Dynamics of fossil fuel neutralization by marine $CaCO_3$, *Global Biogeochem. Cycles,* 12, 259–276.

Azam, F. 1998. Microbial control of oceanic carbon flux: the plot thickens, *Science,* 280, 694–696.

Bates, N.R. 2002. Seasonal variability of the effect of coral reefs on seawater CO_2 and air-sea CO_2 exchange, *Limnol. Ocean,* 47, 43–52.

Behrenfeld, M.J., E. Boss, D.A. Siegel, and D.M. Shea. 2005. Carbon-based ocean productivity and phytoplankton physiology from space, *Global Biogeochem. Cycles,* 19, GB1006, doi:10.1029/2004GB002299.

Behrenfeld, M., R.T. O'Malley, D.A. Siegel, C.R. McClain, J.L. Sarmiento, G.C. Feldman, A.J. Milligan, P.G. Falkowski, R.M. Letelier, and E.S. Boss. 2006. Climate-driven trends in contemporary ocean productivity, *Nature,* 444, 752–755.

Betzer, P.R., R.H. Byrne, J.G. Acker, C.S. Lewis, and R.R. Jolley. 1984. The oceanic carbonate system: a reassessment of biogenic controls, *Science,* 226, 1074–1077.

Bradford-Grieve, J.M., S.D. Nodder, J.B. Jillett, K. Currie, and K.R. Lassey. 2001. Potential contribution that the copepod *Neocalanus tonsus* makes to downward carbon flux in the Southern Ocean, *J. Plankton Res.*, 23, 963–975.

Broecker, W.S., J. Lynch-Stieglitz, E. Clark, I. Hajdas, and G. Bonani. 2001. What caused the atmosphere's CO_2 content to rise during the last 8000 years? *Geochem. Geophys. Geosyst.*, 2, 2001GC000177.

Buitenhuis, E., J. van Bleijswijk, D. Bakker, and M. Veldhuis. 1996. Trends in inorganic and organic carbon in a bloom of *Emiliana huxleyi* in the North Sea, *Mar. Ecol. Prog. Ser.*, 143, 271–282.

Buitenhuis, E.T., H.J.W. de Baar, and M.J.W. Veldhuis. 1999. Photosynthesis and calcification by *Emiliana huxleyi* (Prymnesiophyceae) as a function of inorganic carbon species, *J. Phycol.*, 35, 949–959.

Caldeira, K., D. Archer, J.P. Barry, R.G.J. Bellerby, P.G. Brewer, L. Cao, A.G. Dickson, S.C. Doney, H. Elderfield, V.J. Fabry, R.A. Felly, J.P. Gattuso, P.M. Haugan, O. Hoegh-Guldberg, A.K. Jain, J.A. Kleypas, C. Langdon, J.C. Orr, A. Ridgwell, C.L. Sabine, B.A. Seibel, Y. Shirayama, C. Turley, A.J. Watson, and R.E. Zeebe. 2007. Comment on "Modern-age buildup of CO_2 and its effects on seawater acidity and salinity" by Hugo A. Loaiciga, *Geophys. Res. Lett.*, 34, L18608m doi:10.1029/2096GL027288.

Canadell, J.P., C. Le Quere, M.R. Raupach, C.B. Field, E.T. Buitenhuis, P. Ciais, T.J. Conway, N.P. Gillett, R.A. Houghton, and G. Marland. 2007. Contributions to accelerating atmospheric CO_2 growth from economic activity, carbon intensity, and efficiency of natural sinks, *Proc. Natl. Acad. Sci. USA*, 104, 18866–18870.

Carlson, C.A., H.W. Ducklow, and A.F. Michaels. 1994. Annual flux of dissolved organic carbon from the euphotic zone in the northwestern Sargasso Sea, *Nature*, 371, 405–408.

Carreiro-Silva, M. and T.R. McClanahan. 2001. Echinoid bioerosion and herbivory on Kenyan coral reefs: the role of protection from fishing, *J. Exp Mar. Biol. Ecol.*, 262, 133–153.

Cermeno, P., S. Dutkiewicz, R.P. Harris, M. Follows, O. Schofield, and P.G. Falkowski. 2008. The role of nutricline depth in regulating the ocean carbon cycle, *Proc. Natl. Acad. Sci. USA*, 105, 20344–20349.

Chisholm, J.R.M. 2003. Primary productivity of reef-building crustose coralline algae, *Limnol. Ocean*, 48, 1376–1387.

Chisholm, J.R.M. and D.J. Barnes. 1998. Anomalies in coral reef community metabolism and their potential importance in the reef CO_2 source-sink debate, *Proc. Natl. Acad. Sci. USA*, 95, 6566–6569.

Cook, F.J., W. Hicks, E.A. Gardner, G.D. Carlin, and D.W. Froggatt. 2000. Export of acidity in drainage water from acid sulphate soils, *Mar. Pollut. Bull.*, 41, 319–326.

Doney, S.C. 2006. The dangers of ocean acidification, *Sci. Amer.*, March, 58–65.

Doney, S.C., N. Mahowald, I. Lima, R.A. Feely, F.T. Mackenzie, J.F. Lamarque, and P.J. Rasch. 2007. Impact of anthropogenic atmospheric nitrogen and sulfur deposition on ocean acidification and the inorganic carbon system, *Proc. Natl. Acad. Sci. USA*, 104, 14580–14585.

Dore, J.E., R. Lukas, D.W., Sadler, and D.M. Karl. 2003. Climate-driven changes to the atmospheric CO_2 sink in the subtropical North Pacific Ocean. *Nature*, 424, 754–757.

Feely, R.A., C.L. Sabine, J.M. Hernandez-Ayon, D. Ianson, and B. Hales. 2008. Evidence for upwelling of corrosive "acidified" water onto the continental shelf, *Science*, 320 (5882), 1490–1492.

Feely, R.A., C.L. Sabine, K. Lee, W. Berelson, J. Kleypas, V.J. Fabry, and M.J. Millero. 2004. Impact of anthropogenic CO_2 on the $CaCO_3$ system in the oceans, *Science*, 305 (5682), 362–366.

Fernandez, E., P. Boyd, P.M. Holligan, and D.S. Harbour. 1993. Production of organic and inorganic carbon within a large-scale coccolithophore bloom in the northeast Atlantic Ocean, *Mar. Ecol. Prog. Ser.*, 97, 271–285.

Fletcher, S.E.M., N. Gruber, A.R. Jacobson, S.C. Doney, S. Dutkiewicz, M. Gerber, M. Follows, F. Joos, K. Lindsay, D. Menemenlis, A. Mouchet, S.A. Muller, and J.L. Sarmiento. 2006. Inverse estimates of anthropogenic CO_2 uptake, transport, and storage by the ocean, *Global Biogeochem. Cycles*, 20, GB2002, doi:10.1029/2005GB002530.

Frankignoulle, M., J.P. Gattuso, R. Biondo, I. Bourge, G. Copin-Montegut, and M. Pichon. 1996. Carbon fluxes in coral reefs. II. Eulerian study of inorganic carbon dynamics and measurement of air-sea CO_2 exchanges, *Mar. Ecol. Prog. Ser.*, 145, 123–132.

Friedlingstein, P., P. Cox, R. Betts, L. Bopp, W. von Bloh, V. Brovkin, P. Cadule, S. Doney, M. Eby, I. Fung, G. Bala, H, John, C. Jones, F. Joos, T. Kato, M. Kawamiya, W. Knorr, K. Lindsay, H.D. Matthews, T. Raddatz, P. Rayner, C. Reick, E. Roeckner, K.G. Schnitzler, R. Schnur, K. Strassman, A.J. Weaver, C. Yoshikawa, and N. Zeng. 2006. Climate-carbon cycle feedback analysis: results from the C4MIP model intercomparison, *J. Climate*, 19, 3337–3353.

Gattuso, J.P., D. Allemand, and M. Frankignoulle. 1999. Photosynthesis and calcification at cellular, organismal and community levels in coral reefs: a review of interactions and control by carbonate chemistry, *Amer. Zool.*, 39, 160–183.

Gattuso, J.P., M. Frankignoulle, and S.V. Smith. 1999a. Measurement of community metabolism and significance in the coral reef CO_2 source-sink debate, *Proc. Natl. Acad. Sci. USA*, 96, 13017–13022.

Gattuso, J.P., C.E. Payri, M. Pichon, B. Delesalle, and M. Frankignoulle. 1997. Primary production, calcification, and air-sea CO_2 fluxes of a macroalgal-dominated coral reef community (Moorea, French Polynesia), *J. Phycol.*, 33, 729–738.

Gattuso, J.P., M. Pichon, B. Delesalle, and M. Frankignoulle. 1993. Community metabolism and air-sea CO_2 fluxes in a coral reef ecosystem (Moorea, French Polynesia), *Mar. Ecol. Prog. Ser.*, 96, 259–267.

Gattuso, J.P., M. Pichon, and M. Frankignoulle. 1995. Biological control of air-sea CO_2 fluxes: effect of photosynthetic and calcifying marine organisms and ecosystems, *Mar. Ecol. Prog. Ser.*, 129, 307–312.

Gattuso, J.P., M. Pichon, B. Delesalle, C. Cannon, and M. Frankignoulle. 1996. Carbon fluxes in coral reefs. I. Lagrangian measurement of community metabolism and resulting air-sea CO_2 disequilibrium, *Mar. Ecol. Prog. Ser.*, 145, 109–121.

Gobler, C.J., D.A. Hutchins, N.S. Fisher, E.M. Cosper, and S.A. Sanudo-Wilhelmy. 1997. Release and bioavailability of C, N, P, Se, and Fe following viral lysis of a marine chrysophyte, *Limnol. Ocean*, 42, 1492–1504.

Gruber, N., M. Gloor, S.E.M. Fletcher, S.C. Doney, S. Dutkiewicz, M.J. Follows, M. Gerber, A.R. Jacobson, F. Joos, K. Lindsay, D. Menemenlis, A. Mouchet, S.A. Muller, J.L. Sarmiento, and T. Takahashi.2009. Oceanic sources, sinks, and transport of atmospheric CO_2, *Global Biogeochem. Cycles*, 23, GB1005, doi:10.1029/2008GB003349.

Iglesias-Rodriguez, M.D., P.R. Halloran, R.E.M. Rickaby, I.R. Hall, E. Colmeno-Hidalgo, J.R. Gittins, D.R.H. Green, T. Tyrrell, S.J. Gibbs, P. Von Dassow, E. Rehm, E.V. Armbrust, and K.P. Boessenkool. 2008. Phytoplankton calcification in a high-CO_2 world, *Science*, 320 (5874), 336–340.

Iverson, R.L., W. Esaias, and K. Turpie. 2000. Ocean annual phytoplankton carbon and new production, and annual export production estimated with empirical equations and CZCS data, *Global Change Biol.*, 6, 57–72.

Jin, X., N. Gruber, J.P. Dunne, J.L. Sarmiento, and R.A. Armstrong. 2006. Diagnosing the contribution of phytoplankton functional groups to the production and export of particulate organic carbon, $CaCO_3$, and opal from global nutrient and alkalinity distributions, *Global Biogeochem. Cycles*, 20, GB2015, doi:10.1029/2005GB002532.

Lane, P.V.Z., S.L. Smith, J.L. Urban, and P.E. Biscaye. 1994. Carbon flux and recycling associated with zooplankton pellets on the shelf of the Middle Atlantic Bight, *Deep-Sea Res. II*, 41, 437–457.

Lebrato, M., D. Iglesias-Rodriguez, R.A. Feely, D. Greeley, D.O.B. Jones, N. Suarez-Bosche, R.S. Lampitt, J.E. Cartes, D.R.H. Green, and B. Alker. 2010. Global contribution of echinoderms to the marine carbon cycle: $CaCO_3$ budget and benthic compartments, *Ecol. Mono.*, 80, 441–467.

Lin, C., M. Wood, P. Haskins, T. Ryffel, and J. Lin. 2004. Controls on water acidification and de-oxygenation in an estuarine waterway, eastern Australia, *Estuar. Coast. Shelf Sci.*, 61, 55–63.

Loaiciga, H.A. 2006. Modern-age buildup of CO_2 and its effects on seawater acidity and salinity, *Geophys. Res. Lett.*, 33, L10605, doi:10, 1029/2006GL026305.

Longhurst, A.R., A.W. Bedo, W.G. Harrison, E.J.H. Head, and D.D. Sameoto. 1990. Vertical flux of respiratory carbon by oceanic diel migrant biota, *Deep-Sea Res.*, 37, 685–694.

Maranon, E., M.J. Behrenfeld, N. Gonzalez, B. Mourino, and M.V. Zubkov. 2003. High variability of primary production in oligotrophic waters of the Atlantic Ocean: uncoupling from phytoplankton biomass and size structure. *Mar. Ecol. Prog. Ser.*, 257, 1–11.

Manno, C., V. Tirelli, A. Accornero, and S.F. Uman. 2010. Importance of the contribution of *Limacina helicina* faecal pellets to the carbon pump in Terra Nova Bay (Antarctica), *J. Plankton Res.*, 32, 145–152.

McNeil, B.I. and R.J. Matear. 2006. Projected climate change impact on oceanic acidification, *Carbon Balan. Mgmt.*, 1, 1–2.

Michaels, D. 2010. Airlines find ways to trim fuel use, *Wall Street J.*, March 10, A18.

Migne, A., D. Davoult, and J.P. Gattuso. 1998. Calcium carbonate production of a dense population of the brittle star *Ophiothrix fragilis* (Echinodermata; Ophiuroidea): role in the carbon cycle of a temperate coastal ecosystem, *Mar. Ecol. Prog. Ser.*, 173, 305–308.

Millero, F.J. 2007. The marine inorganic carbon cycle, *Chem. Rev.*, 107, 308–341.

Millero, F.J., K. Lee, and M. Roche. 1998. Distribution of alkalinity in the surface waters of the major oceans, *Mar. Chem.*, 60, 111–130.

Orr, J.C., V.J. Fabry, O. Aumont, L. Bopp, S.C. Doney, R.A. Feely, A. Gnanadesikan, N. Gruber, A. Ishida, F. Joos, R.M. Key, K. Lindsay, E. Maier-Reimer, R. Matear, P. Monfray, A. Mouchet, R.G. Najjar, G.K. Plattner, K.R. Rodgers, C.L. Sabine, J.L. Sarmiento, R. Schlitzer, R.D. Slater, I.J. Totterdell, M.F. Weirig, Y. Yamanaka, and A. Yool. 2005. Anthropogenic ocean acidification over the twenty-first century and its impact on calcifying organisms, *Nature*, 437 (7059), 681–686.

Palacios, S.L. and R.C. Zimmerman. 2007. Response of eelgrass *Zostera marina* to CO_2 enrichment: possible impacts of climate change and potential for remediation of coastal habitats, *Mar. Ecol. Prog. Ser.*, 344, 1–13.

Peyrot-Clausade, M., P. Chabanet, C. Conand, M.F. Fontaine, Y. Letourneur, and M. Harmelin-Vivien. 2000. Sea urchin and fish bioerosion on La Reunion and Moorea reefs, *Bull. Mar. Sci.*, 66, 477–485.

Poulton, A.J., R. Sanders, P.M. Holligan, M.C. Stinchcombe, T.R. Adey, L. Brown, and K. Chamberlain. 2006. Phytoplankton mineralization in the tropical and subtropical Atlantic Ocean, *Global Biogeochem. Cycles*, 20, GB4002, doi:10.1029/2006GB002712.

Randerson, J.T., M.V. Thompson, T.J. Conway, I.Y. Fung, and C.B. Field. 1997. The contribution of terrestrial sources and sinks to trends in the seasonal cycle of atmospheric carbon dioxide, *Global Biogeochem. Cycles*, 11(4), 535–560, doi: 10.1029/97GB02268.

Raupach, M.R., G. Marland, P. Ciais, C. Le Quere, J.G. Canadell, and G. Klepper. 2007. Global and regional drivers of accelerating CO_2 emissions, *Proc. Natl. Acad. Sci. USA*, 104, 10288–10293.

Raven, J., K. Caldeira, H. Elderfield, O. Hoegh-Guldberg, P. Liss, U. Riebesell, J. Shepard, C. Turley, and A. Watson. 2005. Ocean acidification due to increasing atmospheric carbon dioxide, *Policy doc. 12/05, The Royal Society, 6–9 Carlton House Terrace, London SW15AG*, 57 pp.

Raven, J.A. and P.G. Falkowski. 1999. Oceanic sinks for atmospheric CO_2, *Plant Cell Environ.*, 22, 741–755.

Rees, A.P., E. Malcolm, S. Woodward, C. Robinson, D.G. Cummings, G.A. Tarran, and I. Joint. 2002. Size-fractionated nitrogen uptake and carbon fixation during a developing coccolithophore bloom in the North Sea during June 1999, *Deep-Sea Res. II*, 49, 2905–2927.

Ridgwell, A., I. Zondervan, J.C. Hargreaves, J. Bijma, and T.M. Lenton. 2007. Assessing the potential long-term increase of oceanic fossil fuel CO_2 uptake due to CO_2-calcification feedback, *Biogeosciences*, 4, 481–492.

Riebesell, U. 2004. Effects of CO_2 enrichment on marine phytoplankton, *J. Ocean*, 60, 719–729.

Robison, B.H., K.R. Reisenbichler, and R.E. Sherlock. 2005. Giant larvacean houses: rapid carbon transport to the deep sea floor, *Science*, 308. 1609–1611.

Rost, B., I. Zondervan, and D. Wolf-Gladrow. 2008. Sensitivity of phytoplankton to future changes in ocean carbonate chemistry: current knowledge, contradictions and research directions, *Mar. Ecol. Prog. Ser.*, 373, 227–237.

Sabine C., R.A. Feely, N, Gruber, R.M .Key, K. Lee, J.L. Bullister, R. Wanninkhof, C.S. Wong, D.W.R. Wallace, B. Tilbrook, F.J. Millero, T.H. Peng, A. Kozyr, T. Ono, and A.F. Rios. 2004. The oceanic sink for anthropogenic CO_2, *Science*, 305 (5682), 367–371

Sabine, C.L., and R.A. Feely. 2007. The oceanic sink for carbon dioxide. In D. Reay, N. Hewitt, J. Grace, and K. Smith (Eds.). *Greenhouse Gas Sinks* , CABI, Oxfordshire, UK, pp 31–49.

Salisbury, J., M. Green, C. Hunt, and J. Campbell. 2008. Coastal acidification by rivers: a new threat to shellfish?, *Eos Trans. Amer. Geophys. Union*, 89 (50), 513–528.

Sanyal, A., N.G. Hemming, G.N. Hanson, and W.S. Broecker. 1995. Evidence for a higher pH in the glacial ocean from boron isotopes in foraminifera, *Nature* 373, 234–236.

Sarmiento, J.L., J.C. Orr, and U. Siegenthaler. 1992. A perturbation simulation of CO_2 uptake in an ocean general circulation model, *J. Geophys. Res.*, 97, 3621–3645.

Sciandra, A., J. Harlay, D. Lefevre, R. Lemee, P. Rimmelin, M. Denis, and J.P. Gattuso. 2003. Response of coccolithophorid *Emiliana huxleyi* to elevated partial pressure of CO_2 under nitrogen limitation, *Mar. Ecol. Prog. Ser.*, 261, 111–122.

Shitashima, K. 1997. CO_2 supply from deep-sea hydrothermal systems, *Waste Manage.*, 17, 385–390.

Siegenthaler, U., T.R. Stocker, E. Monnin, D. Luthi, J. Schwander, B. Stauffer, D. Raynawed, J.M. Barnoba, M. Fischer, V.L. Delmisto, and J. Jouzel. 2005. Stable carbon cycle-climate relationship during the late Pleistocene, *Science*, 310, 1313–1317.

Takahashi, T. 2004. The fate of industrial carbon dioxide, *Science,* 305 (5682), 352–353.

Takahashi, T., R.A. Feely, R.F. Weiss, R.H. Wanninkhof, D.W. Chipman, S.C. Sutherland, and T.T. Takahashi. 1997. Global air-sea flux of CO_2: an estimate based on measurements of sea-air pCO_2 difference, *Proc. Natl. Acad. Sci. USA*, 94, 8292–8299.

Tans, P. 2009. An accounting of the observed increase in oceanic and atmospheric CO_2 and an outlook for the future, *Oceanography*, 22, 26–35.

Teira, E., M.J. Pazo, M. Quevedo, M.V. Fuentes, F.X. Niell, and E. Fernandez. 2003. Rates of dissolved organic carbon production and bacterial activity in the eastern North Atlantic Subtropical Gyre during summer, *Mar. Ecol. Prog. Ser.*, 249, 53–67.

Tsunogai, S. and S. Noriki. 1991. Particulate fluxes of carbonate and organic carbon in the ocean. Is the marine biological activity working as a sink of the atmospheric carbon? *Tellus*, 43B, 256–266.

Tyrrell, T. 2008. Calcium carbonate cycling in future oceans and its influence on future climates, *J. Plankton Res.*, 30 (2), 141–156.

Vetter, E.W. and C.R. Smith. 2005. Insights into the ecological effects of deep ocean CO_2 enrichment: the impacts of natural CO_2 venting at Loihi seamount on deep sea scavengers, *J. Geophys. Res.*, 110, C09S13, doi:10.1029/2004JC002617.

Walsh, P.J., P. Blackwelder, K.A. Gill, E. Danulat, and T.P. Mommsen. 1991. Carbonate deposits in marine fish intestines: a new source of biomineralization, *Limnol. Ocean*, 36, 1227–1232.

Ware, J.R., S.V. Smith and M.J. Reaka-Kudla. 1991. Coral reefs: sources or sinks of atmospheric CO_2? *Coral Reefs*, 11, 127–130.

White, I., M.D. Melville, B.P. Wilson, and J. Sammut. 1997. Reducing acidic discharges from coastal wetlands in eastern Australia, *Wetlands Ecol. Manage.*, 5, 55–72.

Wilson, R.W., F.J. Millero, J.R. Taylor, P.J. Walsh, V. Christensen, S. Jennings, and M. Grosell. 2009. Contribution of fish to the marine inorganic carbon cycle, *Science*, 323, 359–362.

Zhai, W., M. Dai, W.J. Cai, Y. Wang, and Z. Wang. 2005. High partial pressure of CO_2 and its maintaining mechanism in a subtropical estuary: the Pearl River estuary, China, *Mar. Chem.*, 93, 21–32.

CHAPTER 4

Mode of Action

4.1 General

The behavior of the ocean carbon cycle is continually modified by the increase in atmospheric CO_2 due to fossil fuel combustion and land-use emissions of CO_2 (Andersson et al. 2005). The consequences of a high-CO_2 world and increasing riverine transport of inorganic matter and nutrients arising from human activities were simulated by models between the years 1700 and (projected) 2300. The models show that the global coastal ocean changes from a net source to a net sink of atmospheric CO_2 over time. In the 1700s and the 1800s the direction of the CO_2 flux was from coastal surface waters to the atmosphere, whereas today the net CO_2 flux is into coastal surface waters. These results agree well with recent syntheses of measurements of air-sea CO_2 exchange fluxes from various coastal environments. These models predict that coastal ocean surface water carbonate saturation state would decrease 46% by the year 2100 and 73% by 2300. Observational evidence from the Atlantic Ocean and the Pacific Ocean show that the carbonate saturation state of surface ocean waters has declined during recent decades. For Atolls and other semi-enclosed carbonate systems, the rate of decline is dependent on the residence time of the water in the system. Biogenic production of $CaCO_3$—as based on the positive relation between saturation state and calcification—may decrease as much as 42% by the year 2100 and up to 90% by 2300. If the predicted change in carbonate production were to occur along with rising temperatures, it would threaten the existence of coral reefs and other carbonate systems for some centuries. Cold water carbonate systems are more vulnerable to rising atmospheric

CO_2 than those at lower latitudes. In addition, modelling results predict that carbonate saturation state of coastal sediment pore water will decrease owing to a decrease in pore water pH and increasing CO_2 concentrations attributable to greater deposition and remineralization of organic matter in sediments. In the future, the average composition of carbonate sediments and cements may change as the more soluble magnesium calcites and aragonite are preferentially dissolved and phases of lower solubility, such as calcites with lower magnesium content, increase in percentage abundance in the sediments (Andersson et al. 2005).

Ocean pH is affected by carbon dioxide, dissolution of $CaCO_3$, temperature, depth, and latitude (Millero 2007). The pH of most surface waters in near equilibrium with the atmosphere is 8.1. In the North Atlantic Ocean and the Pacific Ocean, the pH decreases with the addition of CO_2 from the oxidation of organic carbon (Millero 2007). The pH of deep water formed in the North Atlantic Ocean decreases as the waters age and move to the North Pacific Ocean. The pH increase in deep Pacific waters is partly due to the dissolution of $CaCO_3$. The CO_3^{2-} from the dissolution of $CaCO_3$ reacts with H^+ to form HCO_3^-, causing the pH to increase. The pH of deep waters can be as low as 7.5 near 1,000 m. In very deep waters, the pH is affected by the effect of pressure on the ionization of carbonic acid. When the partial pressure of CO_2 increases, the pH decreases. The older intermediate and deep waters in the South Atlantic have a lower pH than the North Atlantic deep waters (Millero 2007). The primary impacts of anthropogenic CO_2 emissions on marine biogeochemical cycles include ocean acidification, global warming-induced shifts in biogeographical provinces, and a negative feedback on atmospheric CO_2 levels by CO_2-fertilized biological production (Oschlies et al. 2008). Predictions of their model include: a negative feedback on atmospheric CO_2 of 34 Gt C by the year 2100 and a 50% increase in the suboxic water volume in response to the respiration of excess organic carbon formed at higher CO_2 levels. This is a significant expansion of marine "dead zones" with severe implications for all higher life forms and for oxygen-sensitive nutrient recycling (Oschlies et al. 2008).

Future changes in ocean acidification caused by emissions of CO_2 to the atmosphere are largely independent of the amounts of climate change (Cao et al. 2007; McNeil and Matear 2007). Climate change—mainly changes in sea-surface temperature and dissolved inorganic carbon concentration—affects predicted changes in ocean pH and calcium carbonate saturation state by a 0.47 reduction in surface ocean pH relative to a pre-industrial value of 8.17, and a reduction in the degree of aragonite saturation from a pre-industrial value of 3.34 to 1.39 by the year 2500 (Cao et al. 2007). With the same CO_2 emissions--but assuming a climate change of +2.5°C—the reduction in projected global mean pH is about 0.48 and the saturation state of aragonite decreases to 1.50. With a climate sensitivity of +4.5°C, these values are 0.51 and 1.61, respectively (Cao et al. 2007). Future projections of surface ocean acidification need only to consider future atmospheric CO_2 levels, not climate change-induced modifications in the ocean (McNeil and Matear 2007).

Global and hemispheric sinks and sources of anthropogenic CO_2 depend on the oceanic sink and the Northern Hemisphere sink (Keeling et al. 1996). Data are consistent with a 1991–1994 period in which the global oceans and the northern land biota each removed the equivalent of 30% of fossil-fuel CO_2 emissions, while tropical land biota were neither strong sources or sinks (Keeling et al. 1996). Measurements of atmospheric oxygen from 1991 to 1997 show that the land biosphere and the world oceans annually sequestered 1.4 and 2.0 gigatons of carbon, respectively (Battle et al. 2000). The rapid storage of carbon by the land biosphere from 1991 to 1997 contrasts with the 1980s, when the land biosphere was neutral. Comparison with measurements of $\delta^{13}CO_2$ implies an isotopic flux of 89 gigatons of carbon annually, in agreement with existing models. Both the $\delta^{13}C$ and the O_2 data show significant interannual variability in carbon storage for the 1991–1997 period, indicating variable carbon uptake by both the land biosphere and the oceans (Battle et al. 2000).

Laboratory studies and field observations continue to show that calcification in corals and coralline algae will decrease in in response to increases in atmospheric CO_2 and that CO_3^{2-} is the limiting component of coral and algal calcification (Kleypas and Langdon 2002). However, long-term trends in calcification from coral bores do not show the expected decrease in calcification since the preindustrial period for reasons that are unclear; further, the biochemical mechanisms for calcification are still poorly understood. Dissolution of high magnesium calcite can potentially buffer the carbonate system in coral reef environments, particularly once seawater becomes undersaturated with that mineral; such dissolution could help maintain saturation states higher than those in the open ocean (Kleypas and Langdon 2002). Dissolution of biogenic materials, however, is confounded by the presence of organic materials, a large range of grain sizes, and fine materials adhering to the surface of larger grains (Bischoff et al. 1987).

4.2 Chemical

The burning of fossil fuels has increased the concentration of CO_2 in the atmosphere from 280 ppm to 385 ppm over the last 200 years(Millero and DiTrolio 2010). This increase is larger than has occurred over the past 800,000 years. Equilibration of increasing amounts of CO_2 with surface waters will decrease the pH of the oceans from a current value of 8.1 to values as low as 7.4 over the next 200 years. Decreasing the pH affects the production of solid $CaCO_3$ by microorganisms in surface waters and its subsequent dissolution. Carbon dioxide dissolution can also affect acid-base equilibria, metal complex formation, solid-liquid equilibria, and the adsorption of ions to charged surfaces (Millero and DiTrolio 2010). Addition of CO_2 to seawater will result in a decrease in pH due to the bicarbonate buffer system in the ocean (Smith and Key 1975; Seibel and Walsh 2003; Dore et al. 2009) according to the following equation:

$$CO_2 + H_2O \rightleftharpoons H_2CO_3 \rightleftharpoons HCO_3^- + H^+ \rightleftharpoons CO_3^{2-} + 2H^+.$$

The concentration of H^+ (in mol/kg seawater) approximates its activity and determines the acidity of the solution. Acidity is commonly expressed on a logarithmic scale as pH:

$$pH = -\log_{10} [H^+]$$

About one third of the carbon dioxide formed from combustion of fossil fuels enters the ocean and reduces the naturally alkaline pH (Doney 2006; Hofmann and Schellnhuber 2009). Carbon dioxide combines with water to form carbonic acid (H_2CO_3), resulting in the release of hydrogen ions (H^+) into solution, leaving both bicarbonate (HCO_3^-) and to a lesser extent carbonate ions (CO_3^{2-}). A small fraction of the carbonic acid remains in solution without dissociating, as does a little carbon dioxide. Some of the bicarbonate ions also dissociate, forming carbonate ions and yet more hydrogen ions. The solubility of calcium carbonate depends on the carbonate ion concentration and thus, indirectly, on pH. The absorption of carbon dioxide has already caused the pH of modern surface water to be about 0.1 unit lower than it was in 1800, with a projected surface ocean decline of an additional 0.3 unit by 2100 from a range of 8.0–8.3 (Doney 2006; Hofmann and Schellnhuber 2009). Rates of change in global CO_2 over the past century are 2 to 3 orders of magnitude higher than most of the changes seen in the past 420,000 years (Hoegh-Guldberg et al. 2007). Biogenic calcification is influenced by the concentration of available carbonate ions, which, in turn, is a decreasing function of increasing CO_2 in the atmosphere (Marubini et al. 2002).

Diminishing pH levels will weaken the ability of calcifying flora and fauna to build their hard parts and will be felt soonest and most severely by organisms that make these parts of aragonite, the form of calcium carbonate that is most prone to dissolution (Doney 2006). Aragonite ($CaCO_3$) is accumulated in coral reefs by calcareous organisms like corals through $CaCO_3$ production (Ohde and Hossain 2004). The saturation degree of seawater ($\Omega_{aragonite}$) is defined as:

$$\Omega = [Ca^{2+}] [CO_3^{2-}]/K'_{sp}$$

where K'sp is the stoichiometric solubility product for aragonite. Corals are assumed to release CO_2 through calcification as follows:

$$Ca^{2+} + 2HCO_3^- \rightleftharpoons CaCO_3 \downarrow + H_2O + CO_2 \uparrow$$

As a result of coral calcification, the $\Omega_{aragonite}$ of seawater is altered through the release of CO_2 in surrounding environments. The elevated CO_2 in marine environments lowers pH through decreasing CO_3^{2-} concentration and leads to a decrease in Ω of seawater (Ohde and Hossain 2004). Ocean uptake of anthropogenic CO_2 alters the seawater chemistry of the global oceans with severe consequences for marine biota (Fabry et al. 2008). Elevated CO_2 is causing the calcium carbonate saturation horizon to shoal in many regions, especially in high latitudes and regions that intersect with pronounced hypoxic zones. The ability of marine organisms, such as pteropod molluscs, foraminiferans, and some benthic invertebrates to produce calcareous skeletal structures is directly affected by seawater CO_2 chemistry. CO_2 also affects marine biota through acid-base imbalance and reduced oxygen transport ability. Ocean acidification coupled with synergistic impacts of other anthropogenic stressors provide great potential for widespread changes to marine ecosystems (Fabry et al. 2008).

Before the Industrial Revolution, most surface waters were substantially oversaturated with respect to aragonite, allowing marine organisms to readily form this mineral. But now, polar surface waters are only marginally oversaturated. Some models predict that these polar waters, particularly those surrounding Antarctica, are expected to become undersaturated resulting in inhibited ability to form aragonite and causing aragonite already formed to dissolve (Doney 2006). Daily aragonite production of two species of subarctic Pacific Ocean molluscan pteropods during July 1985 was 1.8 and 2.6 mg $CaCO_3$ per m^2 daily, respectively (Fabry 1989). Pteropod aragonite accounted for 4 to 13% of the estimated total yearly $CaCO_3$ production of 12–20 g $CaCO_3$ per m^2; however, coccolithophorids are the major producers of $CaCO_3$ at this station, contributing 59 to 77% of the estimated total $CaCO_3$ production (Fabry 1989). The production, transport, and dissolution of carbonate shells is important in the global carbon cycle, affecting the ocean's CO_3^{2-} concentration, which—on glacial-interglacial time scales—is involved in the regulation of atmospheric CO_2 (Jansen et al. 2002). The calcium carbonate saturation state of the ocean suggests that dissolution of calcium carbonate occurs at the sea floor and in the upper part of the water column. With respect to aragonite, dissolution in the water column contributes to the alkalinity maximum in the undersaturated intermediate waters of the North Pacific Ocean—but mechanisms of action need clarification (Jansen et al. 2002). Production of $CaCO_3$ in the world ocean is calculated to be about 5 billion tons (bt) annually, of which about 3 bt accumulate in sediments and the remaining 2 bt is dissolved (Milliman 1993). Nearly half the carbonate in sediments accumulate on reefs, banks, and tropical shelves, and consist largely of metastable aragonite and magnesium calcite. Deep-sea carbonates, mainly from calcitic coccoliths and planktonic foraminifera, have much lower productivity and accumulation rates than does shallow-water carbonates, but they cover a much larger basin area. Twice as much calcium is removed from the oceans by present-day carbonate accumulation as is estimated to be brought in by rivers and hydrothermal activity (1.6 bt), suggesting that outputs have been overestimated or inputs underestimated (Milliman 1993). Ocean pH and calcium carbonate saturation are decreasing due to an influx of anthropogenic CO_2 to the atmosphere (Guinotte et al. 2006). Declining carbonate saturation inhibits the ability of marine organisms to build calcium carbonate skeletons, shells, and tests. The global distribution of deep-sea corals—cold water corals that lack symbiotic algae—as one example, could be limited by the depth of the aragonite saturation horizon in the oceans. (Note: aragonite is the metastable form of calcium carbonate used by scleractinian corals to build their skeletons and the aragonite saturation horizon is the limit between saturated and unsaturated water). More than 95% of deep-sea corals occurred in saturated waters during pre-industrial times, but various models predict that more than 70% of these locations will be in unsaturated waters by the year 2099. Accordingly, seawater chemistry changes need to be carefully monitored (Guinotte et al. 2006).

The change in pelagic calcium carbonate production, as calcite, and dissolution in response to rising atmospheric CO_2 concentrations was modeled (Gehlen et al. 2007). The model predicts values of $CaCO_3$ production and dissolution in line with recent estimates but underestimates $CaCO_3$ dissolution between 0 and 2,000 m—probably because the model only considers calcite. The effect of rising CO_2 on $CaCO_3$ production and dissolution was quantified by model simulations forced with atmospheric CO_2 increasing at a rate of 1% annually from 286 ppm to 1,144 ppm over a period of 140 years. The simulation predicts a decrease of $CaCO_3$ production by 27%. The combined change in production and dissolution of $CaCO_3$ yields an excess uptake of CO_2 from the atmosphere by the ocean of 5.9 Gt C over the period of 140 years. Although this effect is low compared to the total change in dissolved inorganic carbon inventory of about 750 Gt C, it should not hide the potential for major changes in the ecosystem structure (Gehlen et al. 2007). The dissolution of aragonite particles in the ocean mainly depends on the degree of undersaturation of seawater with respect to that mineral (Chung et al. 2004). Most of the upper Atlantic Ocean at depths less than 2,000 m is supersaturated with respect to aragonite whereas much of the deep Atlantic (>2,000 m) is undersaturated. Shallow layers of aragonite-undersaturated water between 20° S and 15° N in the eastern tropical Atlantic are centered at 800 m and are surrounded by aragonite-supersaturated water above and below. Oceanic uptake of anthropogenic CO_2 during the industrial era has caused a significant increase in the size of the undersaturated layers with future expansion likely to occur to the west and south where the degree of supersaturation is low compared to waters of the north. This expansion of the undersaturated aragonite layers is a direct effect of anthropogenic combustion of fossil fuel CO_2 into the ocean (Chung et al. 2004).

Ocean acidification lowers the oceanic saturation states of carbonate minerals and decreases the calcification rates of some marine organisms that provide a range of ecosystem services including coastal protection, aquaculture, and tourism (Cooley et al. 2009). Losses of calcifying organisms or changes in marine food webs could alter global marine harvests which provided 110 million metric tons of food for humans in 2006. By 2050, changes in carbonate mineral saturation rate will be greatest in low-latitude regions with unusual stress on tropical marine ecosystems and societies (Cooley et al. 2009). By the year 2100, atmospheric CO_2 levels could approach 800 ppm and surface water pH could drop from a pre-industrial value of 8.2 to 7.8, increasing the ocean's acidity by about 150% relative to the beginning of the industrial era (Feely et al. 2009). Substantial reductions in surface water carbonate ion concentrations are predicted resulting in aragonite undersaturation or reduction. Model projections indicate that aragonite undersaturation will start to occur in 2020 in the Arctic Ocean and 2050 in the Southern Ocean. By 2050, all of the Arctic will be aragonite-undersaturated. By 2095 all of the Southern Ocean and parts of the North Pacific Ocean will be undersaturated. Also by 2095, most of the Arctic and some parts of the Bering Sea and Chukchi Sea will be undersaturated with respect to calcite, although surface waters from other ocean basins will still be saturated with calcite, but at a level greatly reduced from the present (Feely et al. 2009).

Calcification is a source of CO_2 to the surrounding water and thus a potential source of atmospheric CO_2 due to chemical equilibria involving the CO_2 species (Frankignoulle et al. 1994). Changes in the released CO_2 precipitated carbonate ratio as a function of pCO_2 should be incorporated into future models predicting changes in atmospheric CO_2 content (Frankignoulle et al. 1994). The dissolution process of calcareous sand, however, is not affected by open seawater carbonate chemistry but is controlled by the biogeochemistry of sediment pore water (Leclercq et al. 2002). The incorporation of Sr^{2+} in aragonite and Mg^{2+} in calcite overgrowths are independent of the precipitation rate (Zhong and Mucci 1989). The partition coefficient of strontium in aragonite is about 1.0 and is unaffected by minor salinity variations; however, the magnesium partition coefficient in calcite increases with decreasing salinity, possibly as a result of variations in the sulfate content of the solutions and solids (Zhong and Mucci 1989).

Possible mechanisms for glacial atmospheric CO_2 drawdown and marine carbon sequestration has focused on variable mixing-, equilibration-, or export rates (Skinner 2009). Models now confirm the expectation that a deep sea dominated by an expanded Lower Circumpolar Deep Water (LCDW) water mass holds more CO_2, without any pre-imposed changes in ocean overturning rate, biological export, or ocean-atmosphere exchange. The magnitude of this standing volume effect might be as large as the contributions that have previously been attributed to carbonate compensation, terrestrial biosphere reduction, or ocean fertilization. This standing volume mechanism may help to reduce the amount of glacial-interglacial CO_2 change not explained by other mechanisms (Skinner 2009).

4.3 Physical

Estimates of the atmospheric carbon budget suggest that most of the sink for CO_2 produced by fossil fuel burning and cement production is in the northern hemisphere (Sarmiento et al. 2000). This asymmetry is attributed to a northward preindustrial transport of about 1 Pg C per year in the atmosphere balanced by an equal and opposite southward transport in the ocean. However, the combination of interhemispheric carbon transport, a finite gas exchange, and the biological pump, yield a carbon flux of only 0.12 Pg C per year across the equator to the north. One reason for the low carbon transport is the decoupling of the carbon flux from the interhemispheric heat transport due to the long sea-air equilibration time for the surface CO_2 (Sarmiento et al. 2000).

Sinking rate of phytodetritus in a Norwegian fjord 1,265 m deep was especially rapid (Witte et al. 2003). In that study, freeze-dried *Thalassiosira rotula* equivalent to 1 g organic C/m^2 radiolabeled with ^{13}C were introduced at the surface. After 3 days, all bacteria and benthic macrofauna down to 10 cm sediment depth had taken up ^{13}C from the added phytodetritus, with uptake recorded as soon as 8 h (Witte et al. 2003). The formation of polysaccharide particles during phytoplankton blooms is an important pathway to convert dissolved into particulate organic carbon (Engel et al. 2004). Through an experimental and modelling study, authors state that aggregation processes in the ocean can account for cycling and export of carbon, iron, and thorium (Engel et al. 2004). Rapid vertical transport of particulate matter through the water column may affect sediment oxygen demand at the sea floor, but not always (Sayles et al. 1994). Temporal variations of particulate organic carbon fluxes to the sea floor coupled with limited time for degradation affects sediment oxygen demand by up to four-fold. However, in sediments of the oligotrophic Atlantic Ocean near Bermuda, despite large seasonal variations, measured sediment oxygen consumption does not vary significantly. The contrast could be due to differences in the reactivity of the carbon rain or to differences in biota which might govern response to a pulsed input (Sayles et al. 1994).

In 2005, the upper few hundred meters of the South Atlantic Ocean had higher carbon dioxide concentrations than 1989, and this is consistent with the prevailing view that the sea is taking in carbon dioxide (Doney 2006). These chemical changes cause an upward shift in the saturation horizons for calcite and aragonite, the water levels deep in the sea below which shells of marine organisms made of these minerals dissolve. At present, many deep cold waters are sufficiently acidic to dissolve shells made of calcium carbonate. These ecosystems are undersaturated with respect to calcium carbonate. But shallow, warm surface waters are described as supersaturated in regard to both calcite and aragonite, indicating that these minerals have little tendency to dissolve, The transfer between supersaturated and undersaturated conditions is referred to as the saturation horizon, namely, the level below which calcium carbonate begins to dissolve. The influx of carbon dioxide from the atmosphere has caused the saturation horizons for aragonite and calcite to shift closer to the surface by 50 to 200 meters since 1800. Thus, as the ocean becomes more acidic, the upper shell-friendly portion will become thinner, that is, less of the ocean will be hospitable for calcifying organisms (Doney 2006). Because cold waters are less supersaturated than warm waters for the various forms of calcium carbonates, it appears that ecosystems from high latitudes and deep water may be most vulnerable to oceanic acidification. Doney (2006) predicts that polar surface waters will become undersaturated for aragonite before the end of the century.

4.4 Biological

Future ocean acidification has the potential to adversely affect many species of marine biota, including reduced fertilization success, decreases in larval and adult growth rates, reduced calcification rates, and lowered survival (Melzner et al. 2009). However, certain organisms are less vulnerable to acidification—especially marine ectothermic metazoans with an extensive extracellular fluid volume—as their cells are already exposed to much higher CO_2 values than those of unicellular organisms. A doubling of environmental CO_2 therefore only represents a 10% change in extracellular CO_2 in some marine fishes. High extracellular CO_2 values are related to high metabolic rates and high oxygen consumption. In active metazoans, such as fishes, cephalopods, and many species of crabs, exercise-induced increases in metabolic rate require an efficient ion-regulatory mechanism for CO_2 excretion and acid-base regulation, especially when anaerobic metabolism is involved. These ion transport systems located in gill epithelia, form the basis for compensation of pH disturbances during exposure to elevated environmental CO_2. Compensation of extracellular acid-base status, in turn, may be key to avoiding metabolic depression. So far, maintained performance at high seawater CO_2 has only been observed in adults and juveniles of active high metabolic species with a powerful ion regulatory apparatus. But gametes and early embryonic stages of these tolerant taxa lack specialized ion-regulatory epithelia and are equally sensitive as vulnerable species to elevated seawater CO_2, thus jeopardizing ecological success. More research is needed on these tolerant species to understand the mechanisms of physiological traits that account for differential sensitivities (Melzner et al. 2009).

Elevated carbon dioxide pressures in the water (hypercapnia) produce an acidosis in the blood of organisms that is due to elevated CO_2 in the blood (Burnett 1997). Crustaceans and fish compensate partially for a CO_2-induced acidosis by elevating blood bicarbonate levels. These changes are attributed to ionic exchanges between the blood and the ambient environment, Bivalve molluscs compensate partially for hypercapnia-induced acidosis through elevated hemolymph calcium ion and ammonium concentrations; calcium ions are generated through dissolution of the heavily calcified shell (as quoted in Burnett 1997). Other taxonomic groups respond differently, as indicated below.

4.4.1 Photosynthetic Flora

An important mechanism for the regulation of atmospheric CO_2 concentration is the fixation of CO_2 by marine phytoplankton and the subsequent export of the organically bound carbon to the deeper ocean (Engel et al. 2004a). Different species of algae and macrophytes metabolize carbon in different ways. The inorganic carbon in seawater may be a nutrient limiting the photosynthesis and productivity of certain macroalgae (Holbrook et al. 1988). Decreasing the pH at 2.5 mM inorganic carbon, and thus enhancing the CO_2 by a factor of 30, only slightly increased photosynthesis in 4 of 5 species of marine macroalgae. Authors suggest that bicarbonate, rather than the free CO_2, is the major assimilated form of inorganic carbon (Holbrook et al. 1988).

The evolution of flora capable of oxygenic photosynthesis paralleled a long-term reduction in atmospheric CO_2 and the increase of O_2 (Giordano et al. 2005). The competition between O_2 and CO_2 for active sites became more restrictive to photosynthesis. Many algae and some higher plants acquired mechanisms that use energy to increase the CO_2 concentrations (CO_2 concentrating mechanisms = CCM). Variants are now found among the different groups of algae. The CCMs may be crucial in the energetic and nutritional budgets of a cell and these can be modulated by a multitude of environmental factors (Giordano et al. 2005). Over most of their evolutionary history, marine macroalgae have been subjected to higher mean sea surface temperatures, with rare exposures to surface waters cooler than 5°C (Raven et al. 2002). The Pleistocene glaciations (about 2.5 million years ago) and the preceding cooling during the late Tertiary period provided these low sea surface temperatures at high latitudes, and permitted trans-tropical algal migrations when tropical surface seawater was relatively cool during glaciations. Low sea surface temperatures are predicted to give diffusive CO_2 entry more competitive ability at higher temperatures than flora with elevated CCMs, but the issue remains unresolved (Raven et al. 2002).

Two of the most productive marine calcifying species, the coccolithophores *Coccolithus pelagicus* and *Calcidiscus leptoporus* differ significantly in their responses to changing seawater carbonate chemistry (Langer et al. 2006). In *C. leptoporus*, for example, there is a nonlinear relation of particulate inorganic carbon with increasing CO_2 concentration, with particulate organic carbon remaining constant over the range of CO_2 concentrations tested. In *C. pelagicus* cultures, neither particulate inorganic or particulate organic carbon changed over the range of CO_2 concentrations tested. Authors emphasize the need to consider species-specific effects when evaluating whole ecosystem responses (Langer et al. 2006).

The role of marine phytoplankton in the global carbon cycle is unique because it removes dissolved inorganic carbon from the upper ocean for photosynthesis and redirects it to the deep ocean through sedimentation (Engel 2002). This process is driven mainly by coagulation of single particles into rapidly settling aggregates, enhanced by transparent expolymer particles (TEP). Incubation studies with phytoplankton demonstrate a correlation between CO_2 concentration and the production of TEP, with TEP production linearly related to theoretical CO_2 uptake rates. However, a further increase in atmospheric CO_2 would not lead to a higher rate of dissolved inorganic carbon to TEP conversion since the rate of expolymer carbohydrate production is already at its maximum under the ambient CO_2 concentration (Engel 2002). The role of TEP and dissolved organic carbon for organic carbon partitioning under different CO_2 conditions (*viz.*, 710, 410, and 190 mg/L CO_2) was examined during a mesocosm study with *Emiliana huxleyi* (Engel et al. 2004a). In all mesocosms, TEP concentration increased after nutrient exhaustion and accumulated steadily until the end of the 19-day study. TEP concentration was closely related to *Emiliana* abundance and accounted for an increase in particulate organic carbon concentration of 35% after the onset of nutrient limitation, with production highest at 710 mg/L CO_2. Dissolved organic carbon concentration was not related to *Emiliana* abundance or to TEP concentration. Authors suggest that observed differences between TEP and dissolved organic carbon were determined by different bioavailability, and that the rapid response of a microbial food web may have obscured CO_2 effects on dissolved organic carbon production by autotrophic cells (Engel et al. 2004a).

Data obtained from the photosynthetic gas exchange characteristics of four species of red macroalgae tested support the use of HCO_3^- as a carbon source for *Palmaria palmata* and *Laurencia pinnatifida* (Johnston et al. 1992). However, most of these data support the contention that *Lomentaria articulata* and *Delesseria sanguinea* are restricted to CO_2 as a source of inorganic carbon for photosynthesis (Johnston et al. 1992). Carbon acquisition in relation to CO_2 supply was studied in three bloom-forming macroalgae: a diatom, *Skeletonema costatum*: a flagellate, *Phaeocystis globosa*; and a coccolithophorid, *Emiliana huxleyi* (Rost et al. 2003). Cells were acclimatized to CO_2 levels of 36, 180, 360, and 1,800 mg/L and measured for extracellular and intracellular carbonic anhydrase activity, O_2 evolution, and CO_2 and HCO_3^- uptake rates. Large differences were obtained between species in regard to the efficiency and regulation of carbon acquisition, and other parameters measured. The relative contribution of HCO_3^- to total carbon uptake generally increased with decreasing CO_2, yet strongly differed between species. Authors state that changes in CO_2 availability will influence phytoplankton species succession and distribution (Rost et al. 2003).

Characteristics of inorganic carbon assimilation by photosynthesis were investigated in ten species of brown algae from Arbroath, Scotland (Surif and Raven 1989). All species tested photosynthesized more rapidly at higher external pH values. All had detectable extracellular carbonic anhydrase activity, suggesting that HCO_3^- use could involve catalysis of external CO_2 production. Quantitative differences among the algae examined were noted with respect to inorganic carbon assimilation, especially between those species which were submersed for part or most of the tidal cycle (these species photosynthesized at higher pH values, and had lower CO_2 compensation concentrations) and those which were continually submersed (Surif and Raven 1989).

The role of carbonic anhydrase in marine macroalgae is uncertain. The mechanism of inorganic carbon acquisition by a brown alga *Hizakia fusiforme* is dependent on carbonic anhydrase activity and inhibitors of carbonic anhydrase, such as acetazolamide, significantly depressed the photosynthetic oxygen evolution (Zou et al. 2003). In another study, 16 species of intertidal macroalgae from the Gibralter Strait in southern Spain were examined for external carbonic anhydrase activity and their affinity for inorganic carbon (Mercado et al. 1998). There was no correlation between the ability to use HCO_3^- and the presence of external carbonic anhydrase and the presence of external carbonic anhydrase was not a strong indication for efficient use of HCO_3^- in macroalgae (Mercado et al. 1998).

Rates of cellular uptake of CO_2 and HCO_3^- during steady-state photosynthesis were measured in two diatoms (*Thalassiosira weissflogii, Phaeodactylum tricornutum*) acclimatized to CO_2 partial pressures of 36, 180, 360, and 1,800 mg/L (Burkhardt et al. 2001). Both species responded to diminishing CO_2 supply with an increase of extracellular and intracellular carbonic anhydrase activity, and both took up CO_2 and HCO_3^- simultaneously. In both diatoms the uptake ratios of CO_2/HCO_3^- progressively decreased with decreasing CO_2 concentrations, suggesting that both species of diatoms contain highly efficient and inducible uptake mechanisms for CO_2 and HCO_3^- at concentrations typically encountered in ocean surface waters and both have the ability to adjust uptake rates over a wide range of inorganic carbon supply (Burkhardt et al. 2001). In two species of dinoflagellates, *Amphidinium carterae* and *Heterocapsa oceanica*, exposure to CO_2-induced acidification caused a rapid decrease in external pH from 8 to 7, and a complete suppression of growth after 24 hours (Dason and Colman 2004). Concomitantly with external pH lowering, internal pH of *A. carterae* cells grown at pH 8.0 declined to 7.04, and in *H. oceanica* from 8.14 to 7.22. At lower external pH levels, internal cell pH levels declined to 6.90–7.15. The inability to maintain internal pH probably caused the suppression of growth and the observed loss of photosynthetic activity (Dason and Colman 2004).

Dissolved inorganic carbon (DIC) limitation of photosynthesis brought about by inefficient use of HCO_3^- is a common feature of seagrasses globally (Invers et al. 2001). Photosynthetic inorganic use by four species of seagrasses (*Posidonia oceanica, Cymodocea nodosa, Zostera marina,* and *Phyllospadix torreyi)* was higher at low pH of 6 to 7 than at normal pH of 8.2, demonstrating a clear capacity to use HCO_3^- as an inorganic carbon source for photosynthesis. Thus, photosynthesis of these species was limited by DIC availability in normal seawater(Invers et al. 2001). At normal seawater pH of 8.2, studies with the seagrass *Zostera marina,* showed that HCO_3^- was the major source of inorganic carbon and that bulk CO_2 contributes less than 20% to photosynthesis at that pH (Beer and Rehnberg 1997). Moreover, bicarbonate ion could be acquired via extracellular dehydration to CO_2 when catalyzed by carbonic anhydrase prior to inorganic carbon uptake (Beer and Rehnberg 1997). Subtidal plants of the seagrasses *Cymodocea serrulata* and *Halophila ovalis* had photosynthetic rates that were limited by the ambient inorganic carbon concentration, depending on the irradiance (Schwarz et al. 2000). But intertidal plants of the same species were less dependent on inorganic carbon regardless of irradiance. Both species were able to use HCO_3^- efficiently, and there was stronger evidence for direct uptake of HCO_3^- rather than extracellular dehydration of HCO_3^- to CO_2 prior to inorganic carbon uptake. Photosynthesis was limited by availability of inorganic carbon for the plants growing at depth while those growing in the intertidal zone photosynthesized at close to inorganic carbon saturation (Schwarz et al. 2000).

Photosynthetic performance of the macroalga *Ulva lactuca* was the same at pH 8.2 and pH 5.6 (Drechsler and Beer 1991). Authors conclude that HCO_3^- is transported into cells at defined sites either via facilitated diffusion or active uptake, and that such transport is the basis for elevated CO_2 at the site of ribulose-1,5-bis-phosphate carboxylase/oxygenase carboxylation (Drechsler and Beer 1991). Two mechanisms by which seagrasses use external inorganic carbon include, in addition to uptake of CO_2 formed spontaneously from HCO_3^-: 1/extracellular carbonic anhydrase-mediated conversion of HCO_3^- to CO_2 at normal seawater pH, or in acid zones created by H^+ extrusion, and 2/H^+-driven use of HCO_3^- (Beer et al. 2002). Seagrasses have, until recently, been viewed as having external inorganic carbon utilization systems that are not as efficient as those in macroalgae suggesting that future rises in atmospheric and thus dissolved CO_2 would have a stronger effect on seagrasses than on macroalgae, but these were based on allegedly flawed laboratory studies. Recent studies show that photosynthetic responses of seagrasses and macroalgae to external inorganic carbon are essentially the same and the observed high productivity of seagrass beds are due to efficient H^+- driven mechanisms of HCO_3^- use (Beer et al. 2002).

The relation between incident photon fluence density (PFD) and dissolved inorganic carbon (DIC) availability in three species of rhodophytes (*Gelidium canariensis*, *Gelidium arbuscula*, *Pterocladiella capillacea*) collected from the Canary Islands revealed no detectable carbonic anhydrase activity, high sensitivity to alkaline pH, moderate values of conductance for DIC, and a low capacity for using the external pool of HCO_3^- suggesting that photosynthetic rates were dependent on CO_2 availability (Mercado et al. 2002). The seawater concentration of DIC was insufficient to saturate photosynthesis at high PFD. Photosynthesis is probably limited by incident light rather than DIC despite the apparent low affinity of HCO_3^- in these species (Mercado et al. 2002). In another study, light-limited rhodophytes *Palmaria palmata* and *Laurencia pinnatifida* were not using low concentrations of inorganic carbon as efficiently as plants grown at high PFD (Kubler and Raven 1994). In contrast, growth of *Lomentaria articulata*, which is dependent on diffusive uptake of CO_2 for photosynthesis, was inhibited by the higher PFD (Kubler and Raven 1994).

Coccolithophorids are a group of marine calcifying algae that form extensive blooms in coastal and oceanic regions at temperate and tropical latitudes. These blooms consist mainly of *Emiliana huxleyi*, which is probably the most important calcite producer in the pelagic environment (as quoted in Maranon and Gonzalez 1997). Photosynthesis, calcification, and the patterns of carbon incorporation were measured during a bloom of *Emiliana huxleyi* in the North Sea during summer 1994 (Maranon and Gonzalez 1997). Reduced levels of chlorophyll-*a* and productivity were measured in the coccolithophore-rich waters when compared to stations outside the bloom area. Typical calcification rates within the bloom were 135 mg C/m^2 daily, representing up to 20% of the total carbon incorporation. Relative carbon incorporation into lipids in the bloom area was 1.5 times higher than stations outside the bloom area. Most of the recently-synthesized lipid (70 to 90%) belonged to the neutral lipid fraction. The relation between irradiance and photosynthate partitioning showed that carbon is preferentially channeled into the protein fraction at low light levels, and into lipids at high light levels. Observed patterns of carbon incorporation in *Emiliana* represent a general strategy of energy use found in other groups of marine phytoplankton (Maranon and Gonzalez 1997).

Measurements of surface coccolithophore calcification from the Atlantic Ocean (50°N–50°S) show that correlations between surface calcification, chlorophyll-*a*, and calcite concentrations are statistically significant (Poulton et al. 2007). The contribution of inorganic carbon fixation (calcification) to total carbon fixation (calcification plus photosynthesis) varies between 1 and 10%; a similar contribution for coccolithophores to total organic carbon fixation is usually about the same but sometimes may account for 20% of total carbon fixation in unproductive central subtropical gyres. Composition and turnover times of calcite particles in the upper ocean is probably controlled by interactions between coccolithophore production and detachment rates, species diversity and grazer ecology (Poulton et al. 2007). The relation between inorganic-carbon dependent photosynthesis and calcification was investigated in high- and low-calcifying strains of *Emiliania huxleyi* showing a ten-fold difference in calcification rate (Nimer and Merrett 1992). The high-calcifying cultures showed a four-fold increase in inorganic carbon resulting in 20-fold difference in photosynthetic rate between the two strains at pH 8.3, suggesting that the high photosynthetic rate is sustained by $^{14}CO_2$ released from $H^{14}CO_3$ during calcification. Measurement of bicarbonate transport demonstrated a rapid uptake and achievement of equilibrium between the intracellular and external inorganic carbon concentrations in low- and high-calcifying cells. Subsequent metabolism of the ^{14}C intracellular inorganic carbon pool did not occur in low-calcifying cells suggesting that the block in calcification occurs either in transport into or within the coccolith vesicle (Nimer and Merrett 1992).

Photosynthetic marine cyanobacteria in the genus *Trichodesmium* contribute a large fraction of the new nitrogen entering the oligotrophic oceans (Hutchins et al. 2007; Levitan et al. 2007; Ramos et al. 2007). At projected CO_2 levels for the year 2100 (750 ppm CO_2), N_2 fixation rates of Pacific Ocean and Atlantic Ocean isolates increased 35–100%, and CO_2 fixation rates increased 15–128% relative to present day conditions of 380 ppm CO_2 (Hutchins et al. 2007). Carbon dioxide-mediated rate increases were of similar relative magnitude in both phosphorus-replete and phosphorus-limited cultures, suggesting that this effect may be independent of resource limitation. Neither isolate could grow at 150 mg/L CO_2, but nitrogen and carbon dioxide fixation rate, growth rates, and N:P ratios all increased significantly at 1,500 mg/L CO_2. These parameters were affected only minimally by a 4°C temperature change. The elevated CO_2 levels projected by the end of this century could substantially increase *Trichodesmium* nitrogen and carbon fixation, fundamentally altering the current marine nitrogen and carbon cycles and potentially driving some oceanic regimes towards phosphorus limitation. The relation between marine N_2 fixation and atmospheric CO_2 concentration appears to more complex than previously realized and needs to be considered in the context of the rapidly changing oligotrophic oceans (Hutchins et al. 2007). Acidification of the surface ocean may affect cyanobacterial nitrogen- fixers such as *Trichodesmium* spp. (Levitan et al. 2007). This bloom-forming cyanobacterium was cultured under varying CO_2 levels: 250 mg/L to represent pre-industrial conditions, 400 mg/L (current), and 900 mg/L (future). High CO_2 enhanced nitrogen fixation, filament length, and biomass when compared to ambient and low groups. Photosynthesis and respiration was the same for all groups. Authors suggest that enhanced N_2 fixation and growth in the high CO_2 cultures occurs due to reallocation of energy and resources from carbon concentrating mechanisms required under low and ambient CO_2. Thus, in oceanic regions where light and nutrients are not limiting, projected concentrations of CO_2 are expected to increase nitrogen fixation and growth of *Trichodesmium* and related species, thereby enhancing inputs of nitrogen and increasing primary productivity (Levitan et al. 2007). The effects of 150, 370, or 1,000 mg/L of CO_2 on carbon acquisition by *Trichodesmium* was investigated (Kranz et al. 2009). No differences in growth rates were observed; however, elevated CO_2 levels caused higher carbon and nitrogen quotas and stimulated photosynthesis and N_2 fixation. Minimum extracellular carbonic anhydrase activity was observed, indicating a minor role in carbon acquisition. Rates of CO_2 uptake were small relative to total inorganic fixation, whereas HCO_3^- contributed more than 90% and varied only slightly between CO_2 treatments. Leakage (CO_2 efflux/inorganic carbon uptake) showed pronounced diurnal changes. Authors discount a direct effect of CO_2 on the carboxylation efficiency of ribulose-1,5-biphosphate carboxylase/oxygenase, but point to a shift in resource allocation among photosynthesis, carbon acquisition, and nitrogen fixation under elevated CO_2 levels—with potential biogeochemical implications on productivity stimulation in nitrogen-limited oligotrophic regions and thus provide a negative feedback on rising atmospheric CO_2 levels (Kranz et al. 2009).

Mechanisms for uptake of inorganic carbon for photosynthesis and calcification by two species of symbiont-bearing foraminifera *Amphistegina lobifera* and *Amphisorus hemprichii* were investigated using ^{14}C tracer techniques (ter Kuile et al. 1989). In *A. lobifera*, diffusion is the rate limiting step for total inorganic carbon. Photosynthesis by the isolated symbionts and uptake of CO_3^{2-} for calcification reactions are governed by separate enzymatic reactions. Calcification rates in *A. lobifera* were optimal at calcium concentrations normal for seawater, but were sensitive to inhibitors of respiratory ATP generation and Ca-ATPase. Calcification rates of *A. hemprichii* increased linearly as a function of external inorganic carbon; the dependence of calcification on the CO_3^{2-} concentration was also linear. Calcification in *A. hemprichii* was less sensitive to inhibitors of ATP generation than in *A. lobifera*, suggesting that energy supply is less important for this process in *A. hemprichii* (ter Kuile et al. 1989). The processes of photosynthesis and calcification in *A. lobifera* compete for inorganic carbon (ter Kuile et al. 1989a). Photosynthetic rates were initially high and decreased over time while calcification rates started low and increased over time. Calcification rates were high and constant at all inorganic carbon concentrations when incubated with photosynthesis inhibitors (3(3,4-dichlorophenyl)-1,1-dimethyl-urea). Addition of carbonic anhydrase to the medium, which catalyzes the conversion of HCO_3^- to CO_2, stimulated photosynthesis but inhibited calcification and suggests competition for inorganic carbon between these two processes (ter Kuile et al. 1989a).

Laboratory studies show that increased CO_2 and the accompanying drop in pH is deleterious to all major groups of marine organisms that have hard parts made of calcium carbonate (Doney 2006). Many species of phytoplankton use HCO_3^- for photosynthesis, and because the concentration of bicarbonate ion will remain largely unchanged, these species will benefit according to one scenario (Doney 2006). In laboratory studies with the coccolithophores *Emiliana huxleyi* and *Gephyrocapsa oceanica*, the ratio of particulate inorganic carbon (PIC) to particulate organic carbon (POC) decreased with increasing CO_2 due to both reduced PIC and enhanced POC production (Zondervan et al. 2001). Results were used to formulate a model in which the immediate effect of a decrease in global marine calcification relative to POC production on the potential capacity for oceanic CO_2 uptake was stimulated. Assuming that overall marine biogenic calcification shows a similar response as obtained from coccolithophore studies, the model reveals a negative feedback on increasing atmospheric CO_2 concentrations owing to a decrease in the PIC/POC ratio (Zondervan et al. 2001).

4.4.2 Protists

Food-web processes are important controls of oceanic biogenic carbon flux and ocean-atmosphere CO_2 exchange (Rivkin and Legendre 2001). Two key controlling parameters are the growth efficiencies of the principal trophic components and the rate of carbon remineralization. Since bacterial growth efficiency is an inverse function of temperature, bacterial respiration accounts for most community respiration. This implies that a larger fraction of assimilated carbon is respired at low than at high latitudes and a greater proportion of production can be exported in polar than in tropical regions (Rivkin and Legendre 2001).

4.4.3 Coelenterates

This group is arbitrarily divided into coral reef communities, and all other coelenterates.

4.4.3.1 Coral Reef Communities

The rise in atmospheric CO_2 has caused a decrease in sea surface pH and carbonate ion concentration with adverse effects on calcification in hermatypic corals (Schneider and Erez 2006). Laboratory studies with *Acropora eurystoma* demonstrated that calcification was positively correlated with carbonate ion production. A decrease of about 30% in the carbonate ion concentration (about a 0.2 decrease in pH of seawater) caused a calcification decrease of about 50%. These results suggest that calcification in the oceans today (pCO_2 = 370 ppm) is lower by about 20% compared with pre-industrial time when pCO_2 = 280 ppm. An additional decrease of about 35% is expected if atmospheric CO_2 doubles to 560 ppm. In all experiments, photosynthesis and respiration did not show any significant response to changes in the carbonate chemistry of seawater, suggesting that photosynthesis of symbionts is enhanced by coral calcification at high pH when CO_2 is low. Photosynthesis and calcification support each other mainly through internal pH regulation, which provides carbonate ions for calcification and CO_2 for photosynthesis (Schneider and Erez 2006).

The concentration of CO_2 in the atmosphere is projected to reach twice the preindustrial level by the year 2050, resulting in a decrease of CO_3^{2-} by 30% relative to the preindustrial level and will reduce the calcium carbonate saturation state of the surface ocean by an equal percentage (Langdon et al. 2000). The effect of the projected changes in seawater carbonate chemistry on the calcification of a coral reef mesocosm was investigated over a period of 3.8 years through periodic additions of $NaHCO_3$, Na_2CO_3, and $CaCl_2$. The net community calcification rate responded positively to the ion concentrations of both CO_3^{2-} and Ca^{2+} and the rate was described as a linear function of the ion concentration product $[Ca^{2+}]^{0.69} [CO_3^{2-}]$, suggesting that saturation state is a primary environmental factor influencing coral reef calcification. The pattern was the same for short-term studies of days, or longer-term exposures of months or years, indicating that coral reef organisms were unable to acclimatize to changing saturated states. The predicted decrease in coral reef calcification between the years 1880 and 2065 based on authors' findings is 40% (vs. 14 to 30% for short-term studies of others). This longer study suggests that the impact on coral reefs may be greater than previously suspected and together with projected rising sea level and other anthropogenic stressors, the corals will have difficulty coping (Langdon et al. 2000).

Coral and algal calcification is tightly regulated by the calcium carbonate saturation state of seawater, which is likely to decrease in response to the increase of dissolved CO_2 resulting from the global increase of the partial pressure of atmospheric CO_2 (Langdon et al. 2000; Leclercq et al. 2000). Response of a coral reef community dominated by scleractinian corals, but also including other calcifying biota such as calcareous algae, crustaceans, gastropods, and echinoderms, was investigated using an open-top mesocosm (Leclercq et al. 2000). Community calcification decreased as a function of increasing CO_2 and decreasing aragonite saturation state, as expected. The rate of calcification during the last glacial maximum was about 114% higher than the preindustrial rate. Using the average emission scenario of the Intergovernmental Panel on Climate Change, authors predict that the calcification rate of scleractinian-dominated communities may decrease by

21% between the year 1880 and the time at which atmospheric CO_2 pressure will double in 2065 (Leclercq et al. 2000).

High calcification rates in the scleractinian coral, *Stylophora pistillata*, are due to the symbiotic relationship between coral and photosynthetic dinoflagellates, commonly called zooxanthellae; disruption of this relationship may lead to fatal bleaching (Moya et al. 2008). High concentrations of carbon dioxide induces bleaching (pigmentation loss) in hermatypic corals and in crustose coralline algae through impact on photoprotective mechanisms of the photosystems, although mechanisms of action are imperfectly understood (Anthony et al. 2008). Hermatypic corals calcify about 5 times faster than ahermatypes in the light than in the dark; there is a positive correlation between daily calcium deposition and primary productivity (Goreau et al. 1996). Temperature, light and aragonite saturation state of seawater are important determinants of the global distribution of coral reefs (Marubini et al. 2001). Saturation state of surface seawater is decreasing in response to increases in atmospheric CO_2, causing concern for a global reduction in the rates of reef secretion. Studies with the hermatypic coral *Porites compressa* demonstrated that rising CO_2 will result in a decrease in calcification rate regardless of incident light, indicating that rising CO_2 will impact corals living at all depths (Marubini et al. 2001). Calcification in whole colonies of Mediterranean red coral *Corallium rubrum* was studied by incubating colonies with ^{45}Ca or ^{14}C-aspartic acid and measuring radioisotope uptake in sectioned fractions (Allemand and Benazet-Tambutte 1996). Results suggest that the growth pattern of red coral colonies displays important differences between the tip and the rest of the colony and that secretions from the skeletogenic epithelium govern this process (Allemand and Benazet-Tambutte 1996).

Calcium uptake at the surface of the scleractinian coral *Galaxea fascicularis* involves calcium channels that are coupled directly to inorganic carbon and photosynthesis (Marshall and Clode 2003). These data are consistent with a model suggesting that calcification is a source of CO_2 for photosynthesis in corals (Marshall and Clode 2002). Calcification often results from the high pH of tropical ocean seawater and removal of CO_2 by photosynthesis (Adey 1998). Calcification, in turn, is supportive of photosynthesis by converting seawater HCO_3^- into scarce CO_2. Algae are responsible for most reef accretion of $CaCO_3$, either directly or through photosynthesis-mediated deposition by dinoflagellates in stony coral or by diatoms and chlorophytes in foraminifera. Many algae in reefs are weakly calcified and coated with aragonite crystals. Author concludes that coral reefs are a significant sink globally of ocean alkalinity. Alkalinity carbon is ultimately derived from atmospheric carbon. This is probably not of significance relative to our short-term concern (decades to centuries) for fossil carbon introduction into the earth's atmosphere (Adey 1998).

Inorganic carbon transport and its role in calcification was examined in the coral *Stylophora pistillata* (Furla et al. 2000). For calcification, the major source of dissolved inorganic carbon is metabolic CO_2 (70–75% of total $CaCO_3$ deposition) while only 25–30% originates from the seawater carbon pool, and is dependent on carbonic anhydrase activity. Seawater-dissolved inorganic carbon is transferred from the external medium to the coral skeleton via two pathways, at which time an anion exchanger performs the secretion of dissolved inorganic carbon at the site of calcification. Characterization of the dissolved inorganic carbon supply for symbiotic dinoflagellates demonstrated a dissolved inorganic pool within the tissues. These results confirm the presence of CO_2-concentrating mechanisms in coral cells. The tissue pool is not, however, used as a source for calcification (Furla et al. 2000). But tissues of *Stylophora pistillata* contain calcium-transporting cells and Ca^{2+} is transported across this epithelium via Ca^{2+}-ATPase (Tambutte et al. 1996).

The calcium carbonate in corals, as in shells of other seawater organisms, often comes in two distinct mineral forms: calcite and aragonite (Doney 2006). And some calcite-secreting organisms also add magnesium to the mix. Aragonite and magnesium calcite are more soluble than calcite alone. Thus, corals and pteropods (pelagic gastropod molluscs) which both produce shells of aragonite, and coralline algae which manufacture magnesium calcite, may be especially susceptible to harm from oceanic acidification (Doney 2006) High latitude calcareous phytoplankton and zooplankton with shells formed from calcite, the less soluble form of calcium carbonate, will experience population decline over time. Deep coral communities in the western North Atlantic will also be at risk from water containing high concentrations of carbon from fossil fuel emissions (Doney 2006). Aragonite crystals are localized in the apical cytoplasm of epidermal cells of scleractinian corals, including adults of *Astrangia danae* and planula of *Porites porites* (Hayes and Goreau 1977).

Physiological data of coral calcification indicate that corals use a combination of seawater bicarbonate and respiratory CO_2 for calcification, not seawater carbonate (Jury et al. 2010). To distinguish the effects of pH, carbonate concentration, and bicarbonate concentration on coral calcification, incubations were conducted with the coral *Madracis auretenra* (now *Madracis mirabilis sensu*) in modified seawater chemistries. The corals responded strongly to variation in bicarbonate concentration, but not consistently to carbonate concentration, aragonite saturation state, or pH. Corals calcified at normal or elevated rates under low pH (7.6–7.8) when the seawater bicarbonate concentrations were above 1,800 uM. Conversely, corals incubated at normal pH had low calcification rates if the bicarbonate concentration was lowered. Coral responses to ocean acidification are more diverse than currently thought and authors question the reliability of using carbonate concentration or aragonite saturation state as the sole predictor of ocean acidification on coral calcification (Jury et al. 2010).

Dissolved inorganic carbon uptake by the scleractinian coral *Galaxea fascicularis*, and its delivery to endosymbiotic photosynthetic dinoflagellates (zooxanthellae) suggest that bicarbonate (HCO_3^-) is the main species taken up initially, although zooxanthellae can absorb both CO_2 and HCO_3^- indiscriminately (Goiran et al. 1996). Mechanisms of HCO_3^- uptake as a source of dissolved inorganic carbon for photosynthesis by *Symbiodinium* zooxanthellae in *Galaxea fascicularis* is attributed to two types of DIDS (4,4′-diisothiocyanato-stilbene-2,2′-disulfonic acid) sensitive HCO_3^- carriers, each sharing 50% of the total uptake (Al-Moghrabi et al. 1996). The first is Na^+-dependent and the second is Na^+-independent. The presence of a Na^+-independent Cl^-/HCO_3^- exchange or a Na^+-dependent exchange of Cl^-/HCO_3^- or a Na^+/HCO_3^- is suggested. Authors hypothesize that stimulation of HCO_3^- by the coral host is a consequence of intracellular pH alkalization by zooxanthellae photosynthesis (Al-Moghrabi et al. 1996). Bicarbonate addition to aquaria containing tropical ocean water and branches of *Porites porites* caused a doubling of the skeletal growth rate of the coral (Marubini and Thake 1999).

Increasing additions of $NaHCO_3$ resulted in increasing calcification rate of a *Porites porites* colony in the Barbados until the bicarbonate concentration exceeded three times that of seawater

(Herfort et al. 2008). Photosynthetic rates were also stimulated by HCO_3^- addition but these became saturated at lower concentrations. Similar results were attained with aquarium-acclimated colonies of *Acropora* sp. Calcification rates for *Acropora* were lower in the dark than in the light for a given HCO_3^- concentration but still increased dramatically with HCO_3^- addition, showing that calcification in this coral is light-stimulated but not light-dependent (Herfort et al. 2008).

Photosynthesis in marine free-living algae and macrophytes rely on the concentration of carbon dioxide (Allemand et al. 1998). The major form of dissolved inorganic carbon in seawater is HCO_3^- ions; CO_2 concentration is comparatively low. Because of their electric charge, HCO_3^- ions are poorly diffusible across biological membranes when compared to CO_2. Accordingly, some marine photosynthetic organisms developed CO_2-concentrating mechanisms (CCM) which increase the the CO_2 concentration at the active site of the primary CO_2 fixing enzyme, ribulose-1,5-biphosphate carboxylase-oxygenase, a key enzyme for photosynthesis (Allemand et al. 1998). The main component of the CCM is a pumping mechanism which transports inorganic carbon from the surrounding medium into the chloroplast. Endosymbiotic dinoflagellates of the genus *Symbiodinium* absorb inorganic carbon from the cytoplasm of their host anthozoan cell rather than from seawater (Allemand et al. 1998). *Symbiodinium* spp., generally known as zooxanthellae live within the endodermal cells of their host corals and anemones, separated from the surrounding seawater by the host tissues. The symbiotic association delivers dissolved inorganic carbon to an endodermal site of consumption from an essentially ectodermal site of availability. An HCO_3^--transepithelial active mechanism is present in the host tissues of anthozoans to maintain the photosynthetic rate under saturating irradiance, generating a pH gradient across the epithelium in the process (Allemand et al. 1998).

The shift towards more acidic conditions diminishes the ability of corals and other calcifying organisms to grow by reducing the abundance of carbonate ions in seawater (Doney 2006; Albright et al. 2008; Cohen and Holcomb 2009). Corals living in acidified seawater continue to produce $CaCO_3$ and expend as much energy as corals living in normal seawater to raise the pH of the calcifying fluid; however in acidified seawater, corals are unable to elevate the concentration of carbonate ions to the level required for normal skeletal growth (Cohen and Holcomb 2009). In one study, larvae of the coral *Porites astreoides* were collected from Key Largo, Florida, and subjected to CO_2 levels of 380 mg/L (current; pH 7.99), 560 mg/L (projected for 2065; pH 7.88) and 720 mg/L (2065; pH 7.80); settlement of larvae was unaffected but skeletal extension rate (growth) was reduced about 50% at 560 mg/L and 78% at 720 mg/L (Albright et al. 2008). Although the rate of calcium carbonate production in corals declined with lower pH, the water remains supersaturated with aragonite (Doney 2006). The pH lowering—and thus of carbonate ion concentration—will inhibit the ability of certain taxa to make calcium carbonate. Some of the most abundant life forms that could be affected in this way are the coccolithophorids—which are covered by small plates of calcium carbonate and commonly frequent the ocean surface—and coralline algae that secrete calcium carbonate and contribute to the calcification of many coral reefs (Doney 2006). Coccolithophore production and calcification represent consistent contributions (usually <10%) to organic fixation and total carbon fixation (Poulton et al. 2007). The physiological interactions between photosynthesis and calcification in corals, foraminiferans, and coccolithophores need clarification and resolution would permit better understanding of the effects of elevated temperature and CO_2 on these important biogeochemical systems that currently comprise the bulk of biogenic calcification (Brownlee 2009).

In corals with zooxanthellae the photosynthetic fixation of carbon dioxide and the precipitation of $CaCO_3$ are linked, making it difficult to study carbon transport mechanisms involved only in the calcification process (Tambutte et al. 2007). However, inhibition of the enzyme carbonic anhydrase in *Tubastrea aurea*, a coral lacking zooxanthellae, adversely affects calcification. Organic matrix proteins in *Tubastrea*, which are synthesized by calcifying tissues, play a crucial catalytic role by eliminating the barrier to interconversion of inorganic carbon at the calcification site. The presence of a protein in tissues and in the organic matrix shared common features with prokaryotic carbonic anhydrases (Tambutte et al. 2007). It is unclear as to how, and to what extent, ocean acidification will influence calcium carbonate calcification and dissolution, and affect changes in community structure of present-day coral reefs (Atkinson and Cuet 2008). It is critical to evaluate the extent to which the metabolism of coral reefs is influenced by mineral saturation rates, and to determine a threshold saturation state at which coral communities cease to function as reefs. It is now accepted that dissolved inorganic carbon species are important chemical parameters of the function of key taxa comprising coral reefs, but our understanding is limited and quantification almost non-existent. Recommended research directions include: continue evaluation of effects of carbonate and bicarbonate ions on growth and calcification of key taxa under realistic regimens of light, temperature nutrients, and dissolved inorganic carbon; develop improved models for coral calcification; develop natural coral reef mesocosms to observe whole system changes and community structure competition; expand efforts in monitoring basic CO_2 parameters on a number of coral reefs worldwide, in conjunction with basic community structure data over a period of some decades; and continue studies on the effects of pH on algae and nitrogen-fixing cyanobacteria (Atkinson and Cuet 2008).

Light, temperature, and nutrients, are interrelated and all affect corals. During experimental light-dark cycles, oxygen in tissues of the scleractinian corals *Favia* sp. and *Acropora* sp. reached >250% of air saturation after a few minutes in light (Kuhl et al. 1995). Immediately after darkening, oxygen was rapidly depleted and within 5 min the O_2 concentration at the tissue surface reached <2% of air saturation. The pH of coral tissues changed within 10 min from about 8.5 in the light to 7.3 in the dark. Coral respiration in light was calculated as the difference between the gross and net photosynthesis and was more than 6 times higher at a saturated irradiance than in the dark (Kuhl et al. 1995). Annual variations in skeletal density of large *Porites* colonies from the Great Barrier Reef in Australia over several centuries were strongly and positively correlated with sea surface temperature (Lough and Barnes 1997). A recent decline in calcification of *Porites* from the Great Barrier Reef is tempered by evidence of similar declines and recoveries over the past several hundred years and by evidence that coral calcification on the Great Barrier Reef has been above the long-term average for most of this century and the recent decline may be a return to more normal conditions (Lough and Barnes 1997). High-nutrient, high-CO_2 water (pH 7.5–7.8) of the Waikiki Aquarium in Honolulu, Hawaii, has supported the growth of 57 coral species (Atkinson et al.1995). Inorganic nutrients that are high relative to most natural reef ecosystems include SiO_3, PO_4, NO_3, and NH_4. In contrast, concentrations of organic nutrients (dissolved organic phosphorus and nitrogen) were lower than most tropical surface ocean waters. The coral communities in aquaria took up inorganic nutrients and released organic nutrients. Rates of nutrient uptake into aquaria coral communities were similar to nutrient uptake by natural coral reef communities (Atkinson et al. 1995).

4.4.3.2 Other Coelenterate Species

Carbonic anhydrase activity was detected in 22 species of cnidarians which contain zooxanthellae (Weis et al. 1989). Although carbonic anhydrase has been proposed to function in calcification, its association with zooxanthellae and photosynthetic activity in both calcifying and non-calcifying associations suggests a role in photosynthetic metabolism of algal:cnidarian symbiosis. Carbonic anhydrase acts as a source of CO_2 by releasing CO_2 from bicarbonate, enabling zooxanthellae to maintain high rates of photosynthesis in their intracellular environment (Weis et al. 1989). Photosynthetic rate of the anemone *Aiptasia pulchella* containing symbiotic dinoflagellates increased as dissolved inorganic carbon increased (Weis 1993). Photosynthetic rate decreased dramatically with the addition of acetazolamide, an inhibitor of carbonic anhydrase, the enzyme that catalyzes the interconversion of CO_2 and HCO_3^-. Carbonic anhydrase was localized near the vacuolar membranes surrounding the zooxanthellae (Weis 1993).

The oral epithelial layers of anthozoans have a polarized morphology; photosynthetic symbionts live within endodermal cells facing the coelenteric cavity and are separated from external seawater by the ectodermal layer and the mesoglea (Furla et al. 1998). This morphology affects the supply of inorganic carbon for photosynthesis of the sea anemone *Anemonia viridis* through changes in pH (–0.8 units) and the rate of H^+ fluxes induced by each cell layer of a single anemone tentacle. Polarity plays a significant role both in inorganic carbon absorption and in the control of light-enhanced calcification in cold-water corals (Furla et al. 1998). Symbiotic cnidarians absorb inorganic carbon from seawater to supply intracellular dinoflagellates with CO_2 for their photosynthesis, as judged by studies with ectodermal cells isolated from tentacles of the sea anemone, *Anemonia viridis* (Furla et al.

2000). Authors found that HCO_3^- absorption by ectodermal cells is via H^+ secretion by H^+-ATPase, resulting in the formation of carbonic acid in the surrounding seawater, which is quickly dehydrated into CO_2 by a membrane-bound carbonic anhydrase. Eventually, CO_2 diffuses passively into the cell where it is hydrated in HCO_3^- by a cytosolic carbonic anhydrase (Furla et al. 2000a).

4.4.4 Molluscs

Larval shells of the hardshell clam, *Mercenaria mercenaria*, and the Pacific oyster, *Crassostrea gigas*, both contain amorphous calcium carbonate in addition to aragonite (Weiss et al. 2002). The initially deposited mineral shell is mainly amorphous calcium carbonate that subsequently transforms into aragonite. The post-set juvenile shell and the adult shell of *Mercenaria* also contains aragonite that is less crystalline than non-biogenic aragonite, suggesting that amorphous calcium carbonate is important in larval and adult shell formation (Weiss et al. 2002). Dissolution is a significant source of mortality in juvenile hardshell clams held in sediments undersaturated with carbonate at the sediment-water interface (Green et al. 2004, 2009). Juvenile clams were introduced to sediments undersaturated with respect to aragonite or saturated (controls) in order to evaluate the impact of saturation state and dissolution on survival. Mortality rates were significantly higher in undersaturated sediments, particularly in the smallest clams tested of 0.2 mm shell length. Dissolution-produced mortality may explain, in part, the exponential losses of juvenile bivalves following their transition from the pelagic larval phase to the benthic juvenile phase (Green et al. 2004). When sediments are moderately undersaturated with respect to aragonite, even when overlying water is supersaturated, significant dissolution mortality occurs (Green et al. 2009). Authors recommend additional research on sediment buffering to help preserve commercially and ecologically important sediment-dwelling bivalves in coastal areas (Green et al. 2009).

In the giant clam, *Tridacna gigas*, an inorganic carbon-concentrating mechanism was demonstrated for the symbiotic dinoflagellate *Symbiodinium* sp. (Leggat et al. 1999). Carbon dioxide uptake by *Symbiodinium* sp. from the giant clam was found to support the majority of net photosynthesis (45%–80%) at pH 8.0.

After 2 days, however, HCO_3^- uptake supported 35–95% of net photosynthesis. Intracellular carbon levels were 2 to 7 times higher than would be expected from passive infusion of inorganic carbon. *Symbiodinium* also possess a distinct light-activated intracellular carbonic anhydrase activity, permitting utilization of both CO_2 and HCO_3^- from the medium, suggesting possession of a carbon-concentrating mechanism (Leggat et al. 1999).

Carbon dioxide uptake by the hydrothermal vent clam, *Calyptogena magnifica* and its chemoautotrophic symbionts was demonstrated with gill-symbiont preparations from live clams (Childress et al. 1991b). Clams concentrated sulfide from the medium which was used by the symbionts as a substrate for carbon fixation. Symbionts were shown to be sensitive to sulfide, with inhibition of carbon fixation occurring at low sulfide concnetrations. Incubation media containing sulfide-binding substances were shown to protect the symbionts from this inhibition and to stimulate carbon fixation (Childress et al. 1991b). Biological processes at submarine hydrothermal vents along the mid-oceanic ridge proceed at rates comparable to those from shallow water temperate environments (Lutz et al. 1988). For example, *Calyptogena magnifica* growth at two deep-sea hydrothermal vents in the eastern Pacific Ocean was several orders of magnitude greater than those reported for another bivalve (*Tindaria callistiformis*) from a deep-sea non-vent habitat. Rates of *Calyptogena* shell dissolution decreased markedly with increasing distance (1 to 6 m) from the acidic vent fluids over a period of 210 days (Lutz et al. 1988).

4.4.5 Annelids

First discovered at 2,450 m depth on the Galapagos Rift, a tubeworm, *Riftia pachyptila*, lives symbiotically with intracellular carbon-fixing oxidizing bacteria (Childress et al. 1991a, 1993). Elevated pCO_2 in the worm's environment is a determinant of internal total CO_2 and pCO_2, facilitating CO_2 transport and diffusion to the symbionts. Elevated pCO_2 is essential for this species and for other chemoautotrophic symbioses, most of which occupy environments potentially elevated by pCO_2 (Childress et al. 1991a, 1993). *Riftia pachyptila* is the most conspicuous organism living at deep sea hydrothermal vents along the East Pacific Rise (Goffredi et al. 1997). To support its large size (up to 2.4 m in length) and high growth rates (up to 1.5 m in 2 years), this worm relies exclusively upon internal chemosynthetic bacterial symbionts. The tubeworm must supply inorganic carbon at high rates to the bacteria, which are far removed from the ambient seawater. Acquisition of inorganic carbon is limited by the availability of CO_2, as opposed to bicarbonate. Greatly elevated CO_2 at the vent sites and low environmental pH (as low as 5.6) speed this diffusion. Despite large and variable amounts of total internal CO_2, these worms maintain their extracellular fluid pH stable, and alkaline regardless of the environment. Maintenance of this alkaline pH acts to concentrate inorganic carbon into extracellular fluids. Authors speculate that worms maintain their extracellular pH by active proton-equivalent ion transport via high concentrations of H^+-ATPases. Thus, *Riftia pachyptila* can support the large demand for inorganic carbon by its symbionts owing to the elevated CO_2 in the vent environment, and because it can control extracellular pH in the presence of large inward CO_2 fluxes (Goffredi et al. 1997).

Environmental stressors such as hypercapnia are known to cause acid-base disturbances and lead to metabolic depression in several animals (Reipschlager and Portner 1996). Rate of oxygen consumption in muscle of the marine worm, *Sipunculus nudus*, was measured at various levels of extra- and intracellular pH, CO_2 and HCO_3^-. A reduction of extracellular pH from 7.9 to 7.2 resulted in a significant decrease of intracellular pH to 7.17 and to 7.2 under conditions of hypercapnia with oxygen consumption reduced about 18%. Oxygen consumption was reduced 14% under conditions of hypercapnia when extracellular pH was 7.55 and intracellular pH was 7.32. The depression of aerobic energy turnover in muscle of *S. nudus* is induced by low extracellular pH (Reipschlager and Portner 1996). Further studies with *S. nudus* showed that acidosis caused the ATP demand of Na^+/K^+-ATPase to fall (Portner et al. 2000). This reduction occurred via an inhibiting effect on both Na^+/H^+ and Na^+- dependent $Cl^-/HCO_3^-/Cl^-$ exchange in accordance with a reduction in the ATP demand for acid-base regulation during metabolic depression (Portner et al. 2000).

4.4.6 Arthropods

Intracellular fluid pH in crustaceans regulates the enzymes involved in cell metabolism and accounts for most of the whole animal acid/base equivalents; the extracellular fluid is the intermediary between cellular acid/base production and exchange at the gills and antennal gland (Wheatley and Henry 1992). Compensatory mechanisms include intracellular buffering and transmembrane exchange of acid/base equivalents including an $Na^+/H^+/HCO_3^-/Cl^-$ mechanism and an Na^+/H^+ exchanger. The primary mechanism for acid/base regulation is via electroneutral ion exchangers mainly at the the branchial epithelium and also in renal tubules of species that produce dilute urine. The relation between pH compensation or tolerance and cellular function needs clarification (Wheatley and Henry1992).

Part of the compensatory increase in HCO_3^- that occurs during hypercapnic acidosis in the blue crab, *Callinectes sapidus*, originates from the carbonate reservoirs of the carapace, but is of little quantitative importance either to the total compensatory response or to the maintenance of deposited carbonates; the gills conducted most of the compensatory exchanges with the environment (Cameron 1985). Blue crabs were progressively acclimatized to 2%, 4% and 6% CO_2 with 24 hours at each partial pressure (Cameron and Iwama 1987). Extracellular HCO_3^- and Na^+, K^+, Mg^{2+}, Ca^{2+}, and Cl^- rose during hypercapnic compensation. Percent pH regulation was about 70% at the highest CO_2 level; however, changes in extracellular ionic content did not lead to a measurable change in the concentrations of the major intracellular ions due to superior intracellular buffering (Cameron and Iwama 1987). Exchange of base with the external water was measured in shore crabs, *Carcinus maenas*, under conditions where compensation of blood respiratory acid-base disturbances is known to occur (Truchot 1979). Crabs returned to normal conditions after hypercapnic exposure excrete a significant amount of base to the external water. Conversely, acid output occurs during the first hours of hypercapnic treatment. The main site of these exchange processes is the gills. Exchange mechanisms between the cells and the extracellular fluid are also involved in the compensation of blood respiratory acid-base disturbances (Truchot 1979). Perfusate from isolated gills of the crab, *Carcinus mediterraneus*, held in seawater of pH 8.1 was observed (Lucu and Siebers 1995). Acidification was blocked by acetazolamide and this is attributed to inhibition of the brachial intracellular carbonic anhydrase catalyzing the formation of H^+ ions (Lucu and Siebers 1995).

Oxygen-binding properties of the blood of a deep-sea shrimp, *Glyphocrangon vicaria*, captured at 1,800 m shows no specific effect of CO_2 on oxygen affinity (Arp and Childress 1985).

4.4.7 Elasmobranchs

Spotted dogfish, *Scyliorhinus stellaris*, were exposed to sudden CO_2 increases in inspired seawater (Randall et al. 1976). During hypercapnia, breathing frequency remained constant but gill ventilation increased to 140% of control levels during the first hour, returning to the initial level before normalization of gill ventilation. In contrast, arterial oxygen pressure increased for the entire period of hypercapnia. Hypercapnia resulted in a marked lowering of arterial pH, returning to the initial value in about 5 hours even though hypercapnia was maintained. The rise in arterial pH with little change in external CO_2 was associated with an increase in bicarbonate uptake from the seawater across the gills and was effected by transfer between intracellular tissue compartments and extracellular spaces. This compensatory bicarbonate exchange mechanism in the gills was delayed after onset and after termination of hypercapnia (Randall et al. 1976).

Adults of the Pacific spiny dogfish, *Squalus acanthias*, subjected to acute hypercarbia for 20 min were monitored for arterial blood pressure, systemic vascular resistance, cardiac output and frequency, and ventilatory amplitude and frequency (McKendry et al. 2001). Separate studies were conducted on controls and branchially denervated dogfish to investigate CO_2-chemoreceptive sites on the gills and their link to the autonomic nervous system and cardiorespiratory reflexes. In untreated fish, moderate hypercarbia elicited significant increases in ventilatory amplitude and frequency, and deceases in cardiac frequency and arterial blood pressure with no change in systemic vascular resistance. Denervation of the branchial branches of cranial nerves IX and X to the pseudobranch and each gill arch eliminated all cardiorespiratory responses to hypercarbia. Gill CO_2 chemoreceptors in dogfish are linked to numerous cardiorespiratory reflexes and probably involve cholinergic neurotransmission (McKendry et al. 2001). The effect of hypercapnia on acid-base balance and acid-equivalent transfers were measured in *Squalus acanthias* (Claiborne and

Evans 1992). Cannulated dogfish were maintained in a closed circuit seawater recirculation system for 24 hours of hypercapnic exposure and a subsequent 8 to 24 hour normocapnic recovery period. Respiratory acidosis resulted in a plasma pH depression, which was then almost completely compensated over 24 hours through an increase in plasma bicarbonate. Hypercapnia induced a rapid three-fold increase in HCO_3^- equivalent uptake and also in branchial ammonia (NH_4^+) excretion. These transfers combined for a net H^+ loss to the water, but was reversed during the normocapnic period due to HCO_3^- efflux. Gill Na^+/NH_4^+ exchange contributes to the total ammonia excreted during hypercapnia and NH_3 diffusion predominated during the recovery period. Likewise, Cl^-/HCO_3^- or Cl^-/OH^- exchange may enhance the uptake of HCO_3^- during hypercapnia (Claiborne and Evans 1992).

Although there were no changes in arterial oxygen levels during hypercapnia in skates, *Raja ocellata*, ventilation increased up to 2.7-fold through large changes in stroke volume and small changes in frequency (Graham et al. 1990). There was a decrease in arterial pH which was 65% corrected through HCO_3^- accumulation within 24 h. In contrast, the extradural fluid outside the brain equilibrated slowly and was not involved in ventilatory stimulation. Authors suggest that increased ventilation during hypercapnia may be related to depression in arterial pH (Graham et al. 1990).

4.4.8 Bony Fishes

Mechanisms underlying acid-base transfers across the branchial epithelium of fishes have been studied for more than 70 years (Claiborne et al. 2002). Gill epithelium is the primary site of acid-base transfers to the water via protein transporters of relevant ion transfers. (Edwards et al. 2002). Both Na^+/H^+ exchange and vacuolar-type H^+-ATPase transport H^+ from the fish to the environment. These proteins are localized to mitochondrial-rich cells in the gill epithelium of elasmobranchs and teleosts. A combination of these mechanisms may be used for acid-base regulation in some elasmobranchs. In marine teleosts, an apically located Cl^-/HCO_3^- anion exchanger in chloride cells may be responsible for base excretion. Additional research is recommended on newer molecular approaches on more fish species for further clarification of the roles of various gill membrane transport processes in acid-base balance (Edwards et al. 2002). Acidosis induced by ambient hypercapnia (1% CO_2 in air) in the euryhaline mummichog, *Fundulus heteroclitus*, resulted in an increase in net gill H^+ excretion to the water (Edwards et al. 2005). Net H^+ transfers during acidosis may be driven by different isoforms of Na^+/H^+ transporters (Edwards et al. 2005).

Carbon dioxide excretion and acid-base regulation in fish are linked (Perry and Gilmour 2006), as in other animals, through the reversible reactions of CO_2 and the acid-base equivalents H^+ and HCO_3^-:

$$CO_2 + H_2O \rightleftharpoons H^+ + HCO_3^-$$

These relationships offer two potential routes through which acid-base disturbances may be regulated. Respiratory compensation involves manipulation of ventilation so as to retain CO_2 or enhance CO_2 loss, with eventual readjustment of the CO_2 reaction equilibrium and the resultant changes in H^+ levels. In metabolic compensation, rates of direct H^+ and HCO_3^- exchange with the environment are manipulated to achieve the required regulation of pH; in this case, hydration of CO_2 yields the necessary H^+ and HCO_3^- for exchange. The capacity to use respiratory compensation of the acid-base disturbances is limited and metabolic compensation across the gill is the main mechanism for re-establishing pH balance. The role of the kidney, mechanisms responsible for HCO_3^- reabsorption, and ventilatory responses during acid-base disturbances in this process all need clarification (Perry and Gilmour 2006).

European eels, *Anguilla anguilla*, exposed sequentially to partial pressures of CO_2 in the water ranging from 660 to 10,500 mg/L for 30 minutes at each level, demonstrated a variety of responses including a drop in arterial plasma pH from 7.9 to below 7.2, an increase in arterial CO_2, a progressive decline in arterial blood oxygen content from 10% to <2%, tachycardia, and an increased gill ventilation rate (McKenzie et al. 2002). However, cardiac output and oxygen uptake was unchanged, and mortality was nil. The exceptional tolerance of the European eel to acute hypercapnia is probably a consequence of the tolerance of its heart to acidosis and hypoxia, and a contribution to oxygen uptake for cutaneous respiration (McKenzie et al. 2002). Studies with the olive flounder, *Paralichthys olivaceus*, subjected to seawater of pH 6.18—with pH lowering by either CO_2 or sulfuric acid—showed that all fish within the CO_2 exposure group died in

48 hours, with some deaths after 72 h in the acid group (Hayashi et al. 2004a). Acid-base parameters and plasma ion concentrations were severely perturbed in the CO_2 exposure group, but minimally affected in the acid group. Mortality in the CO_2 group is a direct result of the elevated levels of dissolved CO_2 (50,000 mg/L) and not to the effects of the reduced water pH (Hayashi et al. 2004a). Sturgeons, however, are among the most CO_2 tolerant fishes (Baker et al. 2009). The white sturgeon *Acipenser transmontanus* exposed to elevated concentrations of CO_2 for 48 h exhibited incomplete pH compensation in blood and red blood cells. Despite pH depression of the external environment, intracellular pH of white muscle, heart, brain, and liver did not decrease during a transient blood acidosis. Authors suggest that tight active cellular regulation of internal pH in the absence of external pH compensation is a unique pattern for non-air breathing fishes and may be the basis of CO_2 tolerance (Baker et al. 2009). Significant increases in blood norepinephrine, epinephrine, catecholamines, and cortisol were measured in white sturgeons exposed to acute hypercapnia, and these may form the basis of hypercapnia resistance in this species (Crocker and Cech 1997).

Isolated intestinal segments of the European flounder, *Platichthys flesus*, are capable of active HCO_3^- secretion (Grosell et al. 2005). The HCO_3^- secretion occurs via anion exchange, is dependent on mucosal Cl^-, results in very high mucosal HCO_3^- concentrations, and contributes significantly to Cl^- and fluid absorption. A significant part of the HCO_3^- secretion is fueled by endogenous epithelial CO_2 hydration mediated by carbonic anhydrase. Elevation of serosal HCO_3^- resulted in enhanced HCO_3^- secretion and enhanced Cl^- and fluid absorption. The intestinal epithelium of the European flounder is capable of active HCO_3^- secretion, and may be a general phenomenon of marine teleosts. This secretion provides an explanation for fluid absorption in the presence of net Na^+ absorption (Grosell et al. 2005).

All marine bony fishes produce and excrete carbonate precipitates and may account for as much as 15% of total carbonate production (Wilson et al. 2002; Wilson and Grosell 2003; Grosell and Taylor 2007; Wilson et al. 2009). Carbonate precipitates are excreted by fish via the intestine as result of the osmoregulatory requirement to continuously drink calcium-and magnesium-rich seawater (Wilson et al. 2009). As imbibed seawater passes through the intestine it is alkalinized to pH 8.5–9.2 along with substantial secretion of HCO_3^- ions, typically reaching in gut fluids 20 to 50 times the concentrations found in seawater. These conditions cause precipitation of imbibed Ca^{2+} and some Mg^{2+} ions as insoluble carbonates. Carbonate precipitates formed in the gut are excreted within mucus-coated tubes or pellets or incorporated into feces when fish are feeding. The organic mucus-matrix is rapidly degraded in seawater leaving only inorganic crystals of $CaCO_3$ with high Mg content. There are also acid-base processes involved. The HCO_3^- ions secreted by the intestinal cells into the lumen of fish are derived mainly from metabolic CO_2 reacting with water within intestinal epithelial cells under the catalytic influence of carbonic anhydrase. This reaction produces H^+ which is ultimately excreted into the external seawater via ion-transporting cells in fish gills. Thus, the excretory products include insoluble $CaCO_3$ excreted via the gut, and dissolved H^+ ions excreted via the gills. Solid $CaCO_3$ will rapidly sink and only redissolve at depth, whereas H^+ ions excreted via the gills will remain at the surface ocean. The issue is further confounded by regular vertical migrations of many pelagic fish species, often daily and over several hundred meters (Wilson et al. 2009). Intestinal anion exchange is responsible for high luminal HCO_3^- and CO_3^{2-} concentrations and for elevated intestinal Cl^- (Grosell 2006). The majority of HCO_3^- secreted by the apical anion exchange process is derived from hydration of metabolic CO_2 with the resulting H^+ being extruded via a $Na^+: H^+$ exchange mechanism. High cellular HCO_3^- concentrations are required for the anion exchange process and could be the result of high metabolic activity of the intestinal epithelium and the association of the anion exchange protein with the enzyme carbonic anhydrase (Grosell 2006).

Toadfish, *Opsanus tau*, display an acidosis and decreased HCO_3^- equivalents in blood plasma following transfer to elevated salinity (Genz et al. 2008). Concentrations in plasma of Cl^- and Na^+ increase during the first 24 h at higher salinities, and absorption of Cl^- into the extracellular fluids increasing to a greater extent than Na^+. Increased intestinal HCO_3^- secretion in elevated salinity in toadfish creates an acid-base imbalance that is rapidly corrected by increased H^+ excretion (Genz et al. 2008). The preservation of ion balance and pH despite environmental fluctuations is essential for maintenance of vital cellular functions (Deigweiher et al. 2008). In their study, two members of the HCO_3^- transporter family (Na^+/HCO_3^-; Cl^-/HCO_3^-) are described in gills of the eelpout *Zoarces viviparus* acclimatized to 10,000 mg/L CO_2. Authors demonstrate that rearrangements of Na^+-K^+-ATPase occur under environmental hypercapnia indicating a shift to elevated rates of ion and acid-base regulation (Deigweiher et al. 2008). Isolated perfused gills of two species of Antarctic notothenioids, *Gobionotothen gibberifrons* and *Notothenia corticeps* were subjected to 10,000 mg/L CO_2 under pH compensated hypercapnia (Deigweiher et al. 2010). Energy consumption of *Gobionotothen* rose 100–180% and for *Notothenia* this value was 7 to 56%. High CO_2 concentrations under conditions of compensated acidosis induce cost increments in epithelial processes (Deigweiher et al. 2010). In studies with congrid eels, *Conger conger* subjected for 30 hours to hypercapnia, the arterial pH which had initially declined by 0.4 units was restored by compensatory elevation of plasma bicarbonate concentration (Toews et al. 1983). The continuous rise in bicarbonate is gained by active HCO_3^-/Cl^- ion exchange against the electrochemical gradient between fish and seawater (Toews et al. 1983).

4.4.9 Reptiles

Most reptiles are limited to burst-like hunting behavior that is largely supported through anaerobic metabolism (Ramirez et al. 2007). The ability of reptiles to survive extremes of oxygen availability is dependent on metabolic suppression, tolerance of ionic and pH disturbances, and mechanisms for avoiding free-radical injury during reoxygenation (Bickler and Buck 2007). Turtles are the most anoxia-tolerant vertebrates and meet the challenge of variable oxygen by decreased metabolism, large stores of glycogen, and regulation of ion channel conductance by protein phosphorylation (Bickler and Buck 2007). Hatchlings of the green turtle, *Chelonia mydas*, differ from many other turtle hatchlings by possession of adult hemoglobin in erythrocytes and an additional component with a high affinity for oxygen and marked sensitivity to organic phosphate modulators (Wells and Baldwin 1994).

4.4.10 Birds

Diving birds achieve tolerance to hypoxia by reductions in metabolism, prevention of cellular injury, and the maintenance of functional integrity (Ramirez et al. 2007). Myoglobin is an important oxygen store for supporting aerobic diving in marine birds (Noren et al. 2001). Adult king penguins, *Aptenodytes patagonicus*, contain 3 times more myoglobin in skeletal muscle than juveniles and can remain submerged significantly longer than juveniles. The increase in myoglobin content with age may correspond to increases in activity levels, thermal demands, and time spent in apnea during swimming and diving. The final stage of postnatal development of myoglobin occurs during the initiation of independent foraging (Noren et al. 2001). Ducks (*Aythya* spp.) can dive and remain aerobic for durations that are consistent with their oxygen stores and their metabolic rate (Butler 2004). Ducks, in fact, have a high metabolic rate while diving, mainly because of their positive buoyancy. Larger birds, such as the king penguin and the emperor penguin (*Aptenodytes forsteri*) dive for extended periods beyond their calculated dive limits, yet remain aerobic. Author attributes this to hypometabolism, and hypothermia (Butler 2004).

4.4.11 Mammals

Myoglobin is an important oxygen store for supporting aerobic diving in dolphins, seals, and other marine mammals for activity during submergence (Noren et al. 2001). Myoglobin concentrations in the major locomotor muscles of immature bottlenose dolphins, *Tursiops truncatus*, were only 10% of adult values, thereby preventing prolonged dive durations. Myoglobin concentrations increased significantly during subsequent development. A similar case occurs for northern elephant seals, *Mirounga angustirostris*. The final stage of postnatal development of myoglobin occurs during the initiation of independent foraging, regardless of whether development takes place at sea or on land (Noren et al. 2001).

Fur seals (*Callorhinus* spp.), Weddell seals (*Leptonychotes weddelli*), and probably most cetaceans seem to be able to dive and remain aerobic for durations that are consistent with their elevated stores of usable oxygen, and their metabolic rate while diving being similar to that when they are resting at the surface (Butler 2004). However, grey seals (*Halichoerus grypus*) and elephant seals (*Mirounga* spp.) perform a substantial portion of their dives for periods longer than expected based on these assumptions and yet remain aerobic. This is attributed to reduced metabolic rate below the resting level, and a reduction in body temperature that may contribute to the reduction in metabolic rate (Butler 2004). Hypometabolism and cell protection are not achieved through a single molecular mechanism, but rather through the reconfiguration of all organ systems and networks as well as their underlying cellular and subcellular components (Ramirez et al. 2007).

4.5 Literature Cited

Adey, W.H. 1998. Coral reefs: algal structured and mediated ecosystems in shallow, turbulent, alkaline waters, *J. Phycol.*, 34, 393–406.

Albright, R., B. Mason, and C. Langdon. 2008. Effect of aragonite saturation state on settlement and post-settlement growth of *Porites astreoides* larvae, *Coral Reefs.*, 27, 485–490.

Allemand, D. and S. Benazet-Tambutte. 1998. Dynamics of calcification in the Mediterranean red coral, *Corallium rubrum* (Linnaeus) (Cnidaria, Octocorallia), *J. Exp. Zool.*, 276, 270–278.

Allemand, D., P. Furla, and S. Benazet-Tambutte. 1998. Mechanisms of of carbon acquisition for endosymbiotic photosynthesis in anthozoa, *Canad. J. Bot.*, 76, 925–941.

Al-Moghrabi, S., C. Goiran, D. Allemand, N. Speziale, and J. Jaubert. 1996. Inorganic carbon uptake for photosynthesis by the symbiotic coral–dinoflagellate association. II. Mechanisms for bicarbonate uptake, *J. Exp. Mar. Biol. Ecol.*, 199, 227–248.

Andersson, A.J., F.T. Mackenzie, and A. Lerman. 2005. Coastal ocean and carbonate systems in the high CO_2 world of the Anthropocene, *Amer. J. Sci.*, 305, 875–918.

Anthony, K.R.N, D.J. Kline, G. Diaz-Pulido, S. Dove, and O. Hoegh-Guldberg. 2008. Ocean acidification causes bleaching and productivity loss in coral reef builders, *Proc. Natl. Acad. Sci. USA*, 105, 17442–17446.

Arp, A.J. and J.J. Childress. 1985. Oxygen-binding properties of the blood of the deep-sea shrimp *Glyphocrangon vicaria*, *Physiol. Zool.*, 58, 38–45.

Atkinson, M.J., B. Carlson, and G.L. Crow. 1995. Coral growth in high-nutrient, low-pH seawater: a case study of corals cultured at the Waikiki Aquarium, Honolulu, Hawaii, *Coral Reefs*, 14, 215–223.

Atkinson, M.J. and P. Cuet. 2008. Possible effects of ocean acidification on coral reef biogeochemistry: topics for research, *Mar. Ecol. Prog. Ser.*, 373, 249–256.

Baker, D.W., V. Matey, K.T. Huynh, J.M. Wilson, J.D. Morgan, and C.J. Brauner. 2009. Complete intracellular pH protection during extracellular pH depression is associated with hypercarbia tolerance in white sturgeon,

90   *Oceanic Acidification: A Comprehensive Overview*

Acipenser transmontanus, Amer. J. Physiol., Regul. Integr. Comp. Physiol., 296, R1868–R1880.

Battle, M., M.L. Bender, P.P. Tans, J.W.C. White, J.T. Ellis, T. Conway, and R.J. Francey. 2000. Global carbon sinks and their variability inferred from atmospheric O_2 and $\delta^{13}C$. *Science*, 287, 2467–2470.

Beer, S., M. Bjork, F. Hellblom, and L. Axelsson. 2002. Inorganic carbon utilization in marine angiosperms (seagrasses). *Funct. Plant Biol.*, 29, 349–354.

Beer, S. and J. Rehnberg. 1997. The acquisition of inorganic carbon by the seagrass *Zostera marina*, Aquat. Botany, 56, 277–283.

Bickler, P.E. and L.T. Buck. 2007. Hypoxia tolerance in reptiles, amphibians, and fishes: life with variable oxygen availability, *Ann. Rev. Physiol.*, 69, 145–170.

Bischoff, W.D., F.T. Mackenzie, and F.C. Bishop. 1987. Stabilities of synthetic magnesian calcites in aqueous solution: comparison with biogenic materials, *Geochim. Cosmochim. Acta*, 51, 1413–1423.

Brownlee, C. 2009. pH regulation in symbiotic anemones and corals: a delicate balancing act, *Proc. Natl. Acad. USA*, 106, 16541–16542.

Burkhardt, S., G. Amoroso, U. Riebesell, and D. Sultemeyer. 2001. CO_2 and HCO_3^- uptake in marine diatoms acclimated to different CO_2 concentrations, *Limnol. Ocean*, 46, 1378–1391.

Burnett, L.N. 1997. The challenges of living in hypoxic and hypercapnic aquatic environments, *Amer. Zool.*, 37, 633–640.

Butler, P.J. 2004. Metabolic regulation in diving birds and mammals, *Respir. Physiol. Neurobiol.*, 141, 297–315.

Cameron, J.N. 1985. Compensation of hypercapnic acidosis in the aquatic blue crab, *Callinectes sapidus*: the predominance of external sea water over carapace carbonate as the proton sink, *J. Exp. Biol.*, 114, 197–206.

Cameron, J.N. and G.K. Iwama. 1987. Compensation of progressive hypercapnia in channel catfish and blue crabs, *J. Exp. Biol.*, 133, 183–197.

Cao, L., K. Caldeira, and A.K. Jain. 2007. Effects of carbon dioxide and climate change on ocean acidification and carbonate mineral saturation, *Geophys. Res. Lett.*, 34, L05607, doi:10, 1029/2006GL02865.

Childress, J.J., C.R. Fisher, J.A. Favuzzi, R.E. Kochevar, N.K. Sanders, and A.M. Alayse. 1991a. Sulfide-driven autotrophic balance in the bacterial symbion-containing hydrothermal vent tubeworm, *Riftia pachyptila* Jones, Biol. Bull., 180, 135–153.

Childress, J.J., C.R. Fisher, J.A. Favuzzi, and N.K. Sanders. 1991b. Sulfide and carbon dioxide uptake by the hydrothermal vent clam, *Calyptogena magnifica*, and its chemoautotrophic symbionts, *Physiol. Zool.*, 64, 1444–1470.

Childress, J.J., R.W. Lee, N.K. Sanders, H. Felbeck, D.R. Oros, A. Toulmond, D. Desbruyeres, M. C. Kennicutt, and J. Brooks. 1993. Inorganic carbon uptake in hydrothermal vent tubeworms facilitated by high environmental pCO_2, *Nature*, 362, 147–149.

Chung, S.N., G.H. Park, K. Lee, R.M. Key, F.J. Millero, R.A. Feely, C.L. Sabine, and P.G. Falkowski. 2004. Postindustrial enhancement of aragonite undersaturation in the upper tropical and subtropical Atlantic Ocean: the role of fossil fuel CO_2, *Limnol. Ocean*, 49, 315–321.

Claiborne, J.B., S.L. Edwards, and A. Morrison-Shetlar. 2002. Acid-base regulation in fishes: cellular and modular mechanisms, *J. Exp. Zool.*, 293, 302–319.

Claiborne, J.B. and D.H. Evans. 1992. Acid-base balance and ion transfers in the spiny dogfish (*Squalus acanthias*) during hypercapnia: a role for ammonia excretion, *J. Exp. Zool.*, 261, 9–17.

Cohen, A.L. and M. Holcomb. 2009. Why corals care about ocean acidification. Uncovering the mechanism, *Oceanography*, 22, 118–127.

Cooley, S.R., H.L. Kite-Powell, and S.C. Doney. 2009. Ocean acidification's potential to alter global marine ecosystem services, *Oceanography*, 22, 172–181.

Crocker, C.E. and J.J. Cech, Jr. 1998. Effects of hypercapnia on blood-gas and acid-base status in the white sturgeon, *Acipenser transmontanus*, *J. Comp. Physiol.*, 168, 50–60.

Dason, J.S. and B. Colman. 2004. Inhibition of growth in two dinoflagellates by rapid changes in external pH, *Canad. J. Botany*, 82, 515–520.

Deigweiher, K., T. Hirst, C. Bock, M. Lucassen, and H.O. Portner. 2010. Hypercapnia induced shifts in gill energy budgets of Antarctic notothenioids, *J. Comp. Physiol.*, 180B, 347–359.

Deigweiher, K., N. Koschnick, H.O. Portner, and M. Lucassen. 2008. Acclimation of ion regulatory capacities in gills of marine fish under environmental hypercapnia, *Amer. J. Physiol, Regul. Integr. Comp. Physiol.*, 295. R1660–R1670.

Doney, S.C. 2006. The dangers of ocean acidification, *Sci. Amer.*, March, 58–65.

Dore, J.E., R. Lukas, D.W. Sadler, M.J. Church, and D.M. Karl. 2009. Physical and biogeochemical modulation of ocean acidification in the central North Pacific, *Proc. Natl. Acad. Sci. USA*, 106, 12235–12240.

Drechsler, Z. and S. Beer. 1991. Utilization of inorganic carbon by *Ulva lactuca*, *Plant Physiol.*, 97, 1439–1444.

Edwards, S.L., B.P. Wall, A. Morrison-Shetlar, S. Sligh, J.C. Weakley, and J.B. Claiborne. 2005. The effect of environmental hypercapnia and salinity on the expression of NHE-like isoforms in the gills of a euryhaline fish (*Fundulus heteroclitus*), *J. Exp. Zool.*, 303A, 464–475.

Engel, A. 2002. Direct relationship between CO_2 uptake and transparent exopolymer particles production in natural phytoplankton, *J. Plankton Res.*, 24, 49–53.

Engel, A., B. Delille, S. Jacquet, U. Riebesell, E. Rochelle-Newall, A. Terbruggen, and I. Zondervan. 2004a. Transparent exopolymer particles and dissolved organic carbon by *Emiliana huxleyi* exposed to different CO_2 concentrations: a mesocosm experiment, *Aquat. Microb. Ecol.*, 34, 93–104.

Engel, A., S. Thoms, U. Riebesell, E. Rochelle-Newall, and I. Zondervan. 2004. Polysaccharide aggregation as a potential sink of marine dissolved carbon, *Nature*, 428, 929–932.

Fabry, V.J. 1989. Aragonite production by pteropod molluscs in the subarctic Pacific, *Deep-Sea Res.*, 36, 1735–1751.

Fabry, V.J., B.A. Seibel, R.A. Feely, and J.C. Orr. 2008. Impacts of ocean acidification on marine fauna and ecosystem processes, *ICES J. Mar. Sci.*, 65, 414–432.

Feely, R.A., S.C. Doney, and S.R. Cooley. 2009. Ocean acidification. Present conditions and future changes in a high-CO_2 world, *Oceanography*, 22, 36–47.

Frankignoulle, M., C. Canon, and J.P. Gattuso. 1994. Marine calcification as a source of carbon dioxide: positive feedback of increasing atmospheric CO_2, *Limnol. Ocean*, 39, 458–462.

Furla, P., D. Allemand, and M.N. Orsenigo. 2000a. Involvement of H+-ATPase and carbonic anhydrase in inorganic carbon uptake for endosymbiont photosynthesis, *Amer. J. Physiol. Regul. Integr. Comp. Physiol.*, 278, R870–R881.

Furla, P., S. Benazet-Tambutte, J. Jaubert, and D. Allemand. 1998. Functional polarity of the tentacle of the sea anemone *Anemonia viridis*: role in inorganic carbon acquisition, *Amer. J. Physiol. Regul. Integr. Comp. Physiol.*, 274, R303–R310.

Furla, P., I. Galgani, I. Durand, and D. Allemand. 2000. Sources and mechanisms of inorganic carbon transport for coral calcification and photosynthesis, *J. Exp. Biol.*, 203, 3445–3457.

Gehlen, M., R. Gangsto, B. Schneider, L. Bopp, O. Aumont, and C. Ethe. 2007. The fate of pelagic $CaCO_3$ production in a high CO_2 ocean: a model study, *Biogeosciences*, 4, 505–519.

Genz, J., J.R. Taylor, and M. Grosell. 2008. Effects of salinity on intestinal bicarbonate secretion and compensatory regulation of acid-base balance in *Opsanus tau*, *J. Exp. Biol.*, 211, 2327–2335.

Giordano, M., J. Beardall, and J.A. Raven. 2005. CO_2 concentrating mechanisms in algae: mechanisms, environmental modulation, and evolution, *Ann. Rev. Plant Biol.*, 56, 99–131.

Gledhill, D.K., R. Wanninkhof, F.J. Millero, and M. Eakin. 2008. Ocean acidification of the greater Caribbean region 1996–2996, *J. Geophys. Res.*, 113, C10031, doi:10.1029/2007JC004629.

Goffredi, S.K., J.J. Childress, N.T. Desaulniers, R.W. Lee, F.H. Lallier, and D. Hammond. 1997. Inorganic carbon acquisition by the hydrothermal vent tubeworm *Riftia pachyptila* depends upon high external PCO_2 and upon proton-equivalent ion transport by the worm, *J. Exp. Biol.*, 200, 883–896.

Goiran, C., S. Al-Moghrabi, D. Allemand, and J. Jaubert. 1996. Inorganic carbon uptake for photosynthesis by the symbiotic coral/dinoflagellate association. I. Photosynthetic performances of symbionts and dependence on sea water bicarbonate, *J. Exp. Mar. Biol. Ecol.*, 199, 207–225.

Goreau, T.J., N.I. Goreau, R.K. Trench, and R.L. Hayes. 1996. Calcification rates in corals, *Science*, 274, 117.

Graham, M.G., J.D. Turner, and C.M. Wood. 1990. Control of ventilation in the hypercapnic skate *Raja ocellata*: I. Blood and extradural fluid, *Resp. Physiol.*, 80, 259–277.

Green, M.A., M.E. Jones, C.L. Boudreau, R.L. Moore, and B.A. Westman. 2004. Dissolution mortality of juvenile bivalves in coastal marine deposits, *Limnol. Ocean*, 49, 727–734.

Green, M.A., G.A. Waldbusser, S.L. Reilly, K. Emerson, and S. O'Donnell. 2009. Death by dissolution: sediment saturation state as a mortality factor for juvenile bivalves, *Limnol. Ocean*, 54, 1037–1047.

Grosell, M. 2006. Intestinal anion exchange in marine fish osmoregulation, *J. Exp. Biol.*, 209, 2813–2827.

Grosell, M. and J.R. Taylor. 2007. Intestinal anion exchange in teleost water balance, *Comp. Biochem. Physiol.*, 148A, 14–22.

Grosell, M., C.M. Wood, R.W. Wilson, N.R. Bury, C. Hogstrand, C. Rankin, and F.B. Jensen. 2005. Bicarbonate secretion plays a role in chloride and water absorption of the European flounder intestine, *Amer. J. Physiol. Regul. Integr. Comp. Physiol.*, 288, R936–R946.

Guinotte, J.M., J. Orr, S. Cairns, A. Freiwald, L. Morgan, and R. George. 2006. Will human-induced changes in seawater chemistry alter the distribution of deep-sea scleractinian corals?, *Frontiers Ecol. Environ.*, 4, 141–146.

Hayashi, M., J. Kita, and A. Ishimatsu. 2004. Acid-base responses to lethal aquatic hypercapnia in three marine fishes. *Mar. Biol.*, 144, 153–160.

Hayashi, M., J. Kita, and A. Ishimatsu. 2004a. Comparison of the acid-base responses to CO_2 and acidification in Japanese flounder (*Paralichthys olivaceus*), *Mar. Pollut. Bull.*, 49, 1062–1965.

Hayes, R.L. and N.I. Goreau. 1977. Intracellular crystal-bearing vesicles in the epidermis of scleractinian corals, *Astrangia danae* (Agassiz) and *Porites porites* (Pallas), *Biol. Bull.*, 152, 26–40.

Herfort, L., B. Thake, and I. Taubner. 2008. Bicarbonate stimulation of calcification and photosynthesis in two hermatypic corals, *J. Phycol.*, 44, 91–98.

Hoegh-Guldberg, O., P.J. Mumby, A.J. Hooten, R.S. Steneck, P. Greenfield. E. Gomez, C.D. Harvell, P.F. Sale, A.J. Edwards. K. Caldeira, N. Knowlton, C.M. Eakin, R. Iglesias-Prieto, N. Muthiga, R.H. Bradbury, A. Dubi, and M.E. Hatziolos. 2007. Coral reefs under rapid climate change and ocean acidification, *Science*, 318, 1737–1742.

Hofmann, M. and H.J. Schellnhuber. 2009. Oceanic acidification affects marine carbon pump and triggers extended marine oxygen holes. *Proc. Natl. Acad. Sci. USA*, 106, 3017–3022.

Holbrook, G.P., S. Beer, W.E. Spencer, J.B. Reiskind, J.S. Davis, and G. Bowes. 1988. Photosynthesis in marine macroalgae: evidence for carbon limitation, *Can. J. Bot.*, 66, 577–582.

Hutchins, D.A., F.X. Fu, Y. Zhang, M.E. Warner, Y. Feng, K. Portune, P.W. Bernhardt, and M.R. Mulholland. 2007. CO_2 control of *Trichodesmium* N_2 fixation, photosynthesis, growth rates, and elemental ratios: implications for past, present, and future ocean biogeochemistry, *Limnol. Ocean*, 52, 1292–1304.

Invers, O., R.C. Zimmerman, R.S. Alberte, M. Perez, and J. Romero. 2001. Inorganic carbon sources for seagrass photosynthesis: an experimental evaluation of bicarbonate use in species inhabiting temperate waters, *J. Exp. Biol.*, 265, 203–217.

Jansen, H., R.E. Zeebe, and D.A. Wolf-Gladrow. 2002. Modeling the dissolution of settling $CaCO_3$ in the ocean, *Global Biogeochem. Cycles*, 16, 1027, doi:10.1029/2000GB001279.

Johnston, A.M., S.C. Maberly, and J.A. Raven. 1992. The acquisition of inorganic carbon by four red macroalgae, *Oecologia*, 92, 317–326.

Jury, C.P., R.R. Whitehead, and A.M. Szmant. 2010. Effects of variation in carbonate chemistry on the calcification rates of *Madracis aburetenra* (= *Madracis mirabilis sensu* Wells, 1973): bicarbonate concentrations best predict calcification rates, *Global Change Biol.*, 16, 1632–1644.

Keeling, R.F., S.C. Piper, and M. Heimann. 1996. Global and hemispheric CO_2 sinks deduced from changes in atmospheric O_2 concentration, *Nature*, 381, 218–221.

Kleypas, J. and C. Langdon. 2002. Overview of CO_2-induced changes in seawater chemistry, *Proc. 9th Int. Coral Reef Symp., Bali, Indonesia, 23–27 Oct. 2000*, 2, 1085–1089.

Kranz, S.A., D. Sultemeyer, K.U. Richter, and B. Rost. 2009. Carbon acquisition by *Trichodesmium*: the effect of pCO_2 and diurnal changes, *Limnol. Ocean*, 54, 548–559.

Kubler, J.E. and J.A. Raven. 1994. Consequences of light limitation for carbon acquisition in three rhodophytes, *Mar. Ecol. Prog. Ser.*, 110, 203–209.

Kuhl, M., Y. Cohen, T. Dalsgaard, B.B. Jorgensen, and N.P. Revsbech. 1995. Microenvironment and photosynthesis of zooxanthellae in scleractinian corals studied with microsensors for O_2, pH and light, *Mar. Ecol. Prog. Ser.*, 117, 159–172.

Langdon, C., T. Takahashi, C. Sweeney, D. Chipman, J. Goddard, F. Marubini, H. Aceves, H. Barnett, and M.J. Atkinson. 2000. Effect of calcium carbonate saturation state on the calcification rate of an experimental coral reef, *Global Biogeochem. Cycles*, 14(2), 639–654, doi:19.1029/1999GB001195.

Langer, G., M. Geisen, K.H. Baumann, J. Klas. U. Riebesell, S. Thoms, and J.R. Young. 2006. Species-specific responses of calcifying algae to changing seawater carbonate chemistry, *Geochem. Geophys. Geosyst.* 7. Q09006, doi:10.1029/2005GC001227.

Leclercq, N, J.P. Gattuso, and J. Jaubert. 2000. CO_2 partial pressure controls the calcification rate of a coral community, *Global Change Biol.*, 6, 329–334.

Leggat, W., M.P. Badger, and D. Yellowlees. 1999. Evidence for an inorganic carbon-concentration mechanism in the symbiotic dinoflagellate *Symbiodinium* sp., *Plant Physiol.*, 121, 1247–1255.

Levitan, O., G. Rosenberg, I. Setlik, E. Setlikova, J. Grigel, J. Klepetar, O. Prasil, and I. Berman-Frank. 2007. Elevated CO_2 enhances nitrogen fixation and growth in the marine cyanobacterium *Trichodesmium*, *Global Change Biol.*, 13, 531–538.

Lough, J.M. and D.J. Barnes, 1997. Several centuries of variation in skeletal extension, density and calcification in massive *Porites* colonies from the Great Barrier Reef: a proxy for seawater temperature and a background of variability against which to identify unnatural change, *J. Exp. Mar. Biol. Ecol.*, 211, 29–67.

Lucu, C, and D. Siebers. 1995. Acidification of the gill cells of the shore crab *Carcinus mediterraneus*: its physiological significance, *Helgo. Meers.*, 49, 709–713.

Lutz, R.A., L.W. Fritz, and R.M. Cerrato. 1988. A comparison of bivalve (*Calyptogena magnifica*) growth at two deep-sea hydrothermal vents in the eastern Pacific, *Deep-Sea Res.*, 35, 1793–1810.

Maranon, E. and N. Gonzalez. 1997. Primary production, calcification and macromolecular synthesis in a bloom of the coccolithophore *Emiliania huxleyi* in the North Sea, *Mar. Ecol. Prog. Ser.*, 157, 61–77.

Marshall, A.T. and P.L. Clode. 2002. Effect of increased calcium concentration in sea water on calcification and photosynthesis in the scleractinian coral *Galaxea fascicularis*, *J. Exp. Biol.*, 205, 2107–2113.

Marshall, A.T. and P.L. Clode. 2003. Light-regulated Ca^{2+} uptake and O_2 secretion at the surface of a scleractinian coral *Galaxea fascicularis*, *Comp. Biochem. Physiol. A*, 136, 417–426.

Marubini, F., H. Barnett, C. Langdon, and M.J. Atkinson. 2001. Dependence of calcification on light and carbonate ion concentration for the hermatypic coral *Porites compressa*, *Mar. Ecol. Prog. Ser.*, 220, 153–162.

Marubini, F., C. Ferrier-Pages, and J.P. Cuif. 2002. Suppression of skeletal growth in scleractinian corals by decreasing ambient carbonate-ion concentration: a cross-family comparison, *Proc. Roy. Soc. Lond. B*, 270, 179–184.

Marubini, F. and B. Thake. 1999. Bicarbonate addition promotes coral growth, *Limnol. Ocean*, 44, 716–720.

McKendry, J.E., K.W. Milsom, and S.F. Perry. 2001. Branchial CO_2 receptors and cardiorespiratory adjustments during hypercarbia in Pacific spiny dogfish (*Squalus acanthias*), *J. Exp. Biol.*, 204, 1519–1527.

McKenzie, D.J., E.W. Taylor, A.Z.D. Valle, and J.F. Steffensen. 2002. Tolerance of acute hypercapnic acidosis by the European eel (*Anguilla anguilla*), *J. Comp. Physiol.*, 172B, 339–346.

McNeil, B.I. and R.J. Matear. 2007. Climate change feedbacks on future oceanic acidification, *Tellus*, 59B, 191–198.

Melzner, F., M.A. Gutowska, M. Langenbuch, S. Dupont, M. Lucassen, M.C. Thorndyke, M. Bleich, and H.O. Portner. 2009. Physiological basis for high CO_2 tolerance in marine ectothermic animals: pre-adaptation through lifestyle and ontogeny?, *Biogeosciences*, 6, 2313–2331.

Mercado, J.M., F. Javier, L. Gordillo, F.L. Figueroa, and F.X. Niell. 1998. External carbonic anhydrase and affinity for inorganic carbon in intertidal macroalgae, *J. Exp. Mar. Biol. Ecol.*, 221, 209–220.

Mercado, J.M., F.X. Neill, and M.C. Gil-Rodriguez. 2002. Photosynthesis might be limited by light, not inorganic carbon availability, in three intertidal *Gelidiales* species, *New Phytol.*, 149, 431–439.

Millero, F.J. 2007. The marine inorganic carbon cycle, *Chem. Rev.*, 107, 308–341.

Millero, F.J. and B.R. DiTrolio. 2010. Use of thermodynamics in examining effects of ocean acidification, *Elements*, 6, 299–303.

Milliman, J.D. 1993. Production and accumulation of calcium carbonate in the ocean: budget of a nonsteady state, *Global Biogeochem. Cycles*, 7, 927–957, doi:10.1029/93GB02524.

Moya A., C. Ferrier-Pages, P. Furla, S. Richier, E. Tambutte, D. Allemand, and S. Tambutte. 2008. Calcification and associated physiological parameters during a stress event in the scleractinian coral *Stylophora pistillata*, *Comp. Biochem. Physiol.*, 151A, 29–36.

Nimer, N.A. and M.J. Merrett. 1992. Calcification and utilization of inorganic carbon by the coccolithophorid *Emiliana huxleyi* Lohmann, *New Phytol.*, 121, 173–177.

Noren, S.R., T.M. Williams, D.A. Pabst, W.A. McLellan, and J.L. Dearolf. 2001. The development of diving in marine endotherms: preparing for skeletal muscles of dolphins, penguins, and seals for activity during submergence, *J. Comp. Physiol.*, 171B, 127–134.

Ohde, S. and M.M.M. Hossain. 2004. Effect of $CaCO_3$ (aragonite) saturation state of seawater on calcification of *Porites* coral, *Geochem. J.*, 138, 613–621.

Oschlies, A. K.G. Schulz, U. Riebesell, and A. Schmittner. 2008. Simulated 21st century's increase in oceanic suboxia by CO_2-enhanced biotic carbon export, *Global Biogeochem. Cycles*, 22, GB4008, doi:10.1029/2007GB003147.

Perry, S.F. and K.M. Gilmour. 2006. Acid-base balance and CO_2 excretion in fish: unanswered questions and emerging models, *Resp.* Physiol. *Neurobiol.*, 154, 199–215.

Portner, H.O., C. Bock, and A. Reipschlager. 2000. Modulation of the cost of pHi regulation during metabolic depression: a ^{31}P-NMR study in invertebrate (*Sipunculus nudus*) isolated muscle, *J. Exp. Biol.*, 203, 2417–2428.

Poulton, A.J., T.R. Adey, W.M. Balch, and P.M. Holligan. 2007. Relating coccolithophore calcification rates to phytoplankton community dynamics: regional differences and implications for carbon export, *Deep-Sea Res. II*, 54, 538–557.

Ramirez, J.M., L.P. Folkow, and A.S. Blix. 2007. Hypoxia tolerance in mammals and birds: from the wilderness to the clinic, *Ann. Rev. Physiol.*, 69, 113–143.

Ramos, J.B.E., H. Biswas, K.G. Schulz, J. Laroche, and U. Riebesell. 2007. Effect of rising atmospheric carbon dioxide on the marine nitrogen fixer, *Trichodesmium*, *Global Biogeochem. Cycles*, 21: 10.1029/2006GB002898.

Randall, D.J., N. Heisler, and F. Drees. 1976. Ventilatory response to hypercapnia in the larger spotted dogfish *Scyliorhinus stellaris*, *Amer. J. Physiol.*, 230, 590–594.

Raven, J.A., A.M. Johnston, J.E. Kubler, R. Korb, S.C. Mcinroy, L.L. Handley, C.M. Scrimgeour, D.I. Walker, J. Beardall, M.N. Clayton, M. Vanderklift, S. Fredriksen, and K.H. Dunton. 2002. Seaweeds in cold seas: evolution and carbon acquisition, *Ann. Botany*, 90, 525–536.

Rivkin, R.B. and L. Legendre. 2001. Biogenic carbon cycling in the upper ocean: effects of microbial respiration, *Science*, 291, 2398–2400.

Rost, B., U. Riebesell, S. Burkhardt, and D. Sultmeyer. 2003. Carbon acquisition of bloom-forming marine phytoplankton, *Limnol. Ocean.*, 48, 55–67.

Sarmiento, J.L., P. Monfray, E. Maier-Reimer, O. Aumont, R.J. Murnane, and J.C. Orr. 2000. Sea-air CO_2 fluxes and carbon transport: a comparison of three ocean general circulation models, *Global Biogeochem. Cycles*, 14(4), 1267–1281, doi:10.1029/1999GB900062.

Sayles, F.L., W.R. Martin, and W.G. Deuser. 1994. Response of benthic oxygen demand to particulate organic carbon supply in the deep sea near Bermuda, *Nature*, 371, 686–689.

Schneider, K. and J. Erez. 2006. The effect of carbonate chemistry on calcification and photosynthesis in the hermatypic coral *Acropora eurystoma*, *Limnol. Ocean*, 51, 1284–1293.

Schwarz, A.M., M. Bjork, T. Buluda, M. Mtolera, and S. Beer. 2000. Photosynthetic utilisation of carbon and light by two tropical seagrass species as measured *in situ*, *Mar. Biol.*, 137, 755–761.

Sciandra, A., J. Harlay, D. Lefevre, R. Lemee, P. Rimmelin, M. Denis, and J.P. Gattuso. 2003. Response of coccolithophorid *Emiliana huxleyi* to elevated partial pressure of CO_2 under nitrogen limitation, *Mar. Ecol. Prog. Ser.*, 261, 111–122.

Seibel, B.A. and P.J. Walsh. 2003. Biological impacts of deep-sea carbon dioxide injection inferred from indices of physiological performance, *J. Exp. Biol.*, 206, 641–650.

Skinner, L.C. 2009. Glacial-interglacial atmospheric CO_2 change: a possible "standing volume" effect on deep-ocean carbon sequestration, *Climate of the Past*, 5, 537–550.

Smith, S.V. and G.S. Key. 1975. Carbon dioxide and metabolism in marine environments, *Limnol. Ocean*, 20, 493–495.

Surif, M.B. and J.A. Raven. 1989. Exogenous inorganic carbon sources for photosynthesis in seawater by members of the Fucales and the Laminariales (Phaeophyta): ecological and taxonomic implications, *Oecologia*, 78, 97–105.

Tambutte, E., D. Allemand, E. Mueller, and J. Jaubert. 1996. A compartmental approach to the mechanism of calcification in hermatypic corals, *J. Exp. Biol.*, 199, 1029–1041.

Tambutte, S., E. Tambutte, D. Zoccola, N. Caminiti, S. Lotto, A. Moya, D. Allemand, and J. Adkins. 2007. Characterization and role or carbonic anhydrase on the calcification process of the azooxanthellate coral *Tubastrea aurea*, *Mar. Biol.*, 151, 71–83.

ter Kuile, B., J. Erez, and E. Padan. 1989. Mechanisms for the uptake of inorganic carbon by two species of symbiont-bearing foraminifera, *Mar. Biol.*, 103, 243–251.

ter Kuile, B., J. Erez, and E. Padan. 1989a. Competition for inorganic carbon between photosynthesis and calcification in the symbiont-bearing foraminifer *Amphistegina lobifera*, *Mar. Biol.*, 103, 253–259.

Toews, D.P., G.F. Holeton, and N. Heisler. 1983. Regulation of the acid-base status during environmental hypercapnia in the marine teleost fish *Conger conger*, *J. Exp. Biol.*, 107, 9–20.

Truchot, J.P. 1979. Mechanisms of the compensation of blood and respiratory acid-base disturbances in the shore crab, *Carcinus maenas* (L.), *J. Exp. Zool.*, 210, 407–416.

Weis, V.M. 1993. Effect of dissolved inorganic carbon concentration on the photosynthesis of the symbiotic sea anemone *Aiptasia pulchella* Carlgren: role of carbonic anhydrase, *J. Exp. Biol.*, 174, 209–225.

Weiss, I.M., N. Tuross, L. Addadi, and S. Weiner. 2002. Mollusc larval shell formation: amorphous calcium carbonate is a precursor phase for aragonite, *J. Exp. Zool.*, 293, 478–491.

Wells, R.M. and J. Baldwin. 1994. Oxygen transport in marine green turtle (*Chelonia mydas*) hatchlings: blood viscosity and control of hemoglobin oxygen-affinity, *J. Exp. Biol.*, 188, 103–114.

Wheatley, M.G. and R.P. Henry. 1992. Extracellular and intracellular acid-base regulation in crustaceans, *J. Exp. Zool.*, 263, 127–142.

Wilson, R.W. and M. Grosell. 2003. Intestinal bicarbonate secretion in marine teleost fish—source of bicarbonate, pH sensitivity, and consequences for whole animal acid-base and calcium homeostasis, *Biochim. Biophys. Acta*, 1618, 163–174.

Wilson, R.W., F.J. Millero, J.R. Taylor, P.J. Walsh, V. Christensen, S. Jennings, and M. Grosell. 2009. Contribution of fish to the marine inorganic carbon cycle, *Science*, 323, 359–362.

Wilson, R.W., J.M. Wilson, and M. Grosell. 2002. Intestinal bicarbonate secretion by marine teleost fish—why and how?, *Biochim. Biophys. Acta*, 1566, 182–193.

Witte, U., N. Aberle, M. Sand, and F. Wenzhofer. 2003. Rapid response of a deep-sea benthic community to POM enrichment: an *in situ* experimental study, *Mar. Ecol. Prog. Ser.*, 251, 27–36.

Zhong, S. and A. Mucci. 1989. Calcite and aragonite precipitation from seawater solutions of various salinities: precipitation rates and overgrowth compositions, *Chem. Geol.*, 78, 283–299.

Zondervan, I., R.E. Zeebe, B. Rost, and U. Riebesell. 2001. Decreasing marine biogenic calcification: a negative feedback on rising atmospheric pCO_2, *Global Biogeochem. Cycles*, 15, 507–516.

Zou, D., K. Gao, and J. Xia. 2003. Photosynthetic utilization of inorganic carbon in the economic brown alga, *Hizakia fusiforme*, (sargassaceae) from the South China Sea, *J. Phycol.*, 39, 1095–1100.

CHAPTER 5

Acidification Effects on Biota

5.1 General

Atmospheric carbon dioxide concentration is expected to exceed 500 ppm sometime between 2050 and 2100, a value that significantly exceeds those of at least the past 420,000 years during which most extant marine organisms evolved (Hoegh-Guldberg et al. 2007). The oceans have absorbed nearly half the fossil fuel carbon dioxide emitted into the atmosphere since pre-industrial times causing a measurable reduction in seawater pH and carbonate saturation (Riebesell et al. 2007). By the year 2100, the surface water pH will drop from a pre-industrial value of about 8.2 to about 7.8 (Feely et al. 2009). Various models indicate that aragonite undersaturation will start to occur around the year 2020 in the Arctic Ocean and 2050 in the Southern Ocean; by 2095, most of the Arctic and some parts of the Bering and Chukchi Seas will be undersaturated with respect to calcite (Feely et al. 2009). As levels of dissolved CO_2 in seawater rise, growth rates of calcium-secreting organisms will be reduced due to the effects of decreased ocean CO_3^{2-} concentrations (Feely et al. 2009). If the present rate of fossil fuel combustion continues unabated until the year 2200, a proposed model indicates that the rising atmospheric CO_2 levels may drive global mean temperatures to unprecedented highs, could result in a sea surface pH drop of >0.7 units with inhibition of growth of calcifying organisms, and mayl result in an expansion of hypoxic zones with adverse effects on marine ecosystems (Raven et al. 2005; Hofmann and Schellnhuber 2009).

The first direct impact on humans may be through declining harvests and fishery revenues from shellfish, their predators, and coral reef habitats (Cooley and Doney 2009; UNEP 2010). Substantial revenue declines, job losses, and indirect economic costs may occur if ocean acidification broadly damages marine habitats, alters marine resource availability, and disrupts other ecosystem services. Some developing island and coastal nations that depend heavily on marine and coral ecosystems for food, tourism, and exportable natural resources stand to suffer the most economically. Moreover, coral damage will expose low-lying coastal communities and mangrove ecosystems to storm and wave damage, increasing the potential for economic and social disruption following severe weather events (Cooley and Doney 2009; UNEP 2010).

The oceans have taken up about one-third of the total CO_2 released into the atmosphere by human activities over the past 200 years (Fabry 2008). The addition of CO_2 to the surface ocean changes seawater chemistry causing a decrease in pH and carbonate ion concentration, and an increase in the concentrations of bicarbonate ion and hydrogen ion. Ocean absorption of anthropogenic CO_2 also reduces the saturation state of seawater with respect to calcite and aragonite, two common types of calcium carbonate secreted by marine biota. Studies with calcareous organisms indicate that calcification is strongly dependent on the carbonate saturation state of seawater, suggesting that ocean acidification will adversely impact calcifying taxa. Evidence is now accumulating on the acidifying effects of CO_2 in seawater and its consequences for marine calcifiers (Fabry 2008). New tools are being developed, such as functional genomics, that measure gene expression in larvae of stony corals and sea urchins in an attempt to reveal how calcifying marine organisms will respond to ocean acidification (Hofmann et al. 2008). There is much variability among marine calcifiers to CO_2-induced ocean acidification (Ries et al. 2009). Results of 60-day laboratory studies with 18 species of benthic marine biota—crustaceans, coelenterates, echinoderms, rhodophytes, chlorophytes, gastropods and bivalve molluscs, and annelids–including organisms producing aragonite, low Mg-calcite, and high-Mg calcite forms of $CaCO_3$, demonstrated that 10 of the 18 species studied had reduced rates of net calcification or dissolution. However, in seven species, net calcification increased under intermediate and high levels of pCO_2, and one species showed no response. These varied responses may reflect differences among organisms in ability to regulate pH at the site of calcification, in extent to which outer shell layer is protected by an organic covering, in solubility of their shell or skeletal mineral, and in extent to which they use photosynthesis. Specific mechanisms involved are unknown, but results suggest that effects of elevated atmospheric pCO_2 on marine calcification are more varied than previously thought (Ries et al. 2009).

Carbon dioxide elicits acidosis not only in the water but also in tissues and body fluids of marine fauna (Portner et al. 2004). Sensitivity is maximal in ommastrephid squid, which are characterized by a high metabolic rate and extremely pH-sensitive blood oxygen transport. Most fish are less sensitive than invertebrates, owing to intracellular blood pigments and higher capacities of fish to compensate for CO_2-induced acid-base disturbances; however, effects of hypercapnia on deep-sea fishes are largely unknown. Sensitivity to CO_2 is hypothesized to be related to the animal's energy requirements and mode of life, with long-term adverse effects predicted on growth, reproduction, and survival (Portner et al. 2004).

Knutzen (1981) in an early review of decreased pH levels on marine biota states that organisms can tolerate a pH decrease of 0.5 to 1.0 units for extended periods. However, at lower pH values, effects include: reduced photosynthesis, reduced growth, reduced calcification, and greater susceptibility to trace metals in algae; reduced growth in bacteria; reduced growth, survival, egg development, pumping frequency, and abnormal heart beat in bivalve molluscs; and increased death rate in crustaceans (Knutzen 1981). Laboratory studies demonstrate that declining pH negatively impacts calcification in corals, molluscs, crustaceans, echinoderms, coralline algae, and phytoplankton (Kurihara 2008; Wootton et al. 2008). Ocean acidification has negative impacts on the fertilization, cleavage, larva, settlement and reproductive stages of several marine calcifiers, although there are species-specific differences in tolerance to the high CO_2 levels (Kurihara 2008). Laboratory studies also demonstrate that CO_2-induced marine acidification adversely affects performance at the level of reproduction, behavior and growth, especially in the lower marine invertebrates which have a low capacity to compensate for disturbances in extracellular ion and acid-base status; however, in the case of CO_2-induced oceanic hypercapnia and acidification, effects may be so small that evidence for changes in the field are incomplete (Portner 2008). Development of a cause and effect understanding is needed beyond empirical observations for a more accurate projection of ecosystem effects and for quantitative scenarios. Identification of the mechanisms through which CO_2-related oceanic physicochemistry affect organism fitness, survival, and well-being is crucial with this research strategy (Portner 2008).

5.2 Photosynthetic Flora

Oceanic primary production of biomass is important in the global carbon cycle because it constitutes 40% of total primary production on Earth (Hein and Sand-Jensen 1997). In the nutrient-poor central Atlantic Ocean, primary production increased 15 to 19% in response to elevated CO_2 concentrations which simulate the CO_2 rise in surface waters predicted to occur over the next 100 to 200 years. Although the overall response of oceanic primary production to the rise in CO_2 would be relatively small, the influence on species composition of phytoplankton assemblages could be profound, depending on the kinetics of carbon use (Hein and Sand-Jensen 1997). Different species of coastal flora differ significantly in their response to pH (Hinga 2002). The optimum pH for growth ranges from 6.3 to 10.0 under laboratory conditions; however, some species grow well over a wide range of pH, while others show best growth over a 0.5–1.0 pH unit change. Growth rates may vary between clones of the same species as well as different seawater regimens (Hinga 2002). To fully discern the effect of ocean acidification on calcifying phytoplankton a combination of in-water (ship, autonomous vehicle, buoy), optical and chemical measurements, plus satellite optical measurements are needed to measure standing stock of particulate inorganic carbon across ocean basins, and over seasonal to decadal time scales (Balch and Fabry 2008).

5.2.1 Crustose Coralline Algae

Ocean acidification inhibits calcification rates, recruitment rates, and growth rates in many species of crustose coralline algae (Kuffner et al. 2008; Kleypas and Yates 2009). Coralline algae produce large quantities of high-magnesium calcite, the most soluble form of the common marine carbonate minerals (Kleypas and Yates 2009). This group is extremely widespread in the ocean and is found in equatorial as well as polar regions. The broad distribution suggests that coralline algae have the evolutionary capacity to adapt to a wide range of conditions. But individual reef-building species seem particularly vulnerable to dissolution under conditions of increasing CO_2 concentrations with no demonstrable ability to adapt. When CO_2 levels were elevated by 365 mg/L over present-day levels, calcification rates of some species decreased by as much as 250% (net dissolution) and successful recruitment was diminished (Kleypas and Yates 2009). Elevated concentrations of CO_2 caused bleaching of the crustose coralline alga *Porolithon onkodes*, leading to negative productivity and high rates of dissolution. This sensitive reef building species may be pushed beyond their threshold within the next few decades should CO_2 concentrations exceed 520–700 mg/L or seawater pH declines to 7.85–7.95 (Anthony et al. 2008). Calcification in the coralline alga *Corallina pilulifera* at 350 mg CO_2/L (pH 8.2) displayed diurnal variation, with faster rates during the light than the dark period (Gao et al. 1993a). Addition of 1,600 mg CO_2/L (pH 7.6) inhibited calcification due to the decreased pH resulting from CO_2 addition (Gao et al. 1993a). In the coralline red alga *Amphiroa anceps* photosynthesis was highest between pH 6.5 and 7.5 (Borowitzka 1981). At pH 9 to 10 there is still a significant photosynthetic rate, suggesting that this alga can use HCO_3^- as a substrate for photosynthesis. At pH 8.5 and 8.8, calcification rate continues to increase with increasing concentration of total inorganic carbon. Between pH 7 and 9, calcification rate of another species of coralline red alga *Amphiroa foliacea* is proportional to the photosynthetic rate. At all pH values examined, the calcification rate of living *A. foliacea* plants in the dark and of dead plants is proportional to the CO_3^{2-} ion concentration, suggesting little metabolic involvement in calcification processes in the dark, whereas calcification in live *A. foliacea* in the light is influenced by photosynthesis rate as well as the CO_3^{2-} ion concentration in the medium (Borowitzka 1981).

Effects of elevated partial pressure of CO_2 and temperature, alone and in combination, on survival, calcification, and dissolution were measured in the crustose coralline alga *Lithophyllum cabiochae* (Martin and Gattuso 2009). Algae were maintained for one year at ambient conditions of irradiance, at ambient or elevated temperature (+3°C) and at ambient (400 ppm) or elevated CO_2 (700 ppm). Algal necroses appeared at elevated temperature first at 700 ppm CO_2 and then at 400 ppm CO_2; algal death occurred only under elevated temperature. Net calcification decreased by 50% when both temperature and CO_2 were elevated, but no effect was found under elevated temperature or elevated CO_2 alone. The dissolution of dead algal thalli was 2 to 4 times more rapid at elevated CO_2. Authors suggest that net dissolution is likely to exceed net calcification in *L. cabiochae* by the end of this century, with major consequences for biodiversity and biogeochemistry in coralligenous communities dominated by these algae (Martin and Gattuso 2009).

5.2.2 Coccolithophores

Coccolithophores are major calcium bicarbonate producers in the world's oceans, accounting for about 33% of the total marine $CaCO_3$ production (Iglesias- Rodriguez et al. 2008). *Emiliana huxleyi* is a coccolithophorid alga distributed worldwide, and usually blooms in temperate regions after the spring diatom bloom. This species produces coccoliths, or platelets, made of $CaCO_3$ in the mineral form of calcite (Buitenhuis et al. 1999). A study with *Emiliania huxleyi* shows that calcification and net primary production are significantly increased by high CO_2 (Iglesias-Rodriguez et al. 2008). Field evidence from the deep ocean is consistent with these laboratory results, indicating that over the past 220 years there has been a 40% increase in average coccolith mass that will probably continue to respond to rising atmospheric CO_2 levels (Iglesias-Rodriguez et al. 2008). Based on increased oceanic CO_2 of 750 mg/L, and single species culture experiments at high latitudes, authors estimate that coccolithophores will double their rate of calcification and photosynthesis. Another study with coccolithophores showed different results. Reduced calcite production at increased CO_2 concentrations is reported in two species of coccolithophorids: *Emiliania huxleyi* and *Gephyrocapsa oceanica* (Riebesell et al. 2000). Also reported is an increase in malformed coccoliths and incomplete coccospheres. Diminished calcification resulted in a reduction in the ratio of calcite precipitation to organic matter production. In that study, the carbonate system of the growth medium was adjusted by adding acid or base to cover a a range from pre-industrial CO_2 levels (280 ppm) to about three times pre-industrial levels (about 750 ppm).

Over this range, the two species tested showed a slight increase in photosynthetic carbon fixation of 8.5% (*E. huxleyi*) and 18.6% (*G. oceanica*). The ratio of calcite to particulate organic matter production (calcite/POC) for the two species decreased by 21.0% (*E. huxleyi*) and 52.5% (*G. oceanica*) when CO_2 levels increased from 280 to 750 ppm. Similar results were obtained during incubation of plankton from the north Pacific Ocean when exposed to experimentally elevated CO_2 levels (Riebesell et al. 2000). It is important to state that light intensity in this study was always below 150 *u*mol $m^{-2} s^{-1}$ (Riebesell et al. 2000).

Direct effects of CO_2 and related changes in seawater carbonate chemistry on marine planktonic organisms via outdoor mesocosms was studied over a 19-day period (Engel et al. 2005). Mesocosms (11 m^3 capacity) were subjected to 190, 410, or 710 mg/L CO_2, and fertilized with nitrate and phosphate. The resultant blooms were dominated by the coccolithophorid *Emiliana huxleyi* in all mesocosms. But high CO_2 levels affected growth rates, calcification rates, elemental uptake, and accumulation and loss processes. Changes in CO_2 directly affect cell physiology with likely effects on the marine biogeochemistry (Engel et al. 2005). In another mesocosm study wherein *Emiliana huxleyi* was the dominant bloom organism, primary production and calcification were studied in response to three different CO_2 levels: glacial, present, and projected for the year 2010 (Delille et al. 2005). A comparison of year 2010 with glacial CO_2 showed the following: no change in net community productivity; a delay in the onset of calcification by 24–48 hours, a reduction in the duration of the calcifying phase during the bloom; a 40% decrease of net community calcification; and an enhanced loss of organic carbon from the water column. Collectively, these results suggest a shift in the ratio of organic carbon to calcium carbonate production and vertical flux with rising atmospheric CO_2 (Delille et al. 2005).

In general, increased absorption of CO_2 as a result of anthropogenic CO_2 release will result in decreased calcification by coccolithophores (Riebesell et al. 2000; Zondervan et al. 2001; Bijma et al. 2002; Riebesell 2004; Delille et al. 2005). However, different coccolithophore species show different calcification responses (Iglesias-Rodriguez et al. 2008). For *Emiliana huxleyi* and *Gephyrocapsa oceanica*, increased CO_2 causes a decrease in calcification (Riebesell et al. 2000; Zondervan et al. 2001; Delille et al. 2005), negligible calcification with *Coccolithus pelagicus* (Langer et al. 2006), and an increase followed by a decrease for *Calcidiscus leptorus* (Langer et al. 2006).

Changes to seawater inorganic carbon and nutrient concentrations in response to CO_2 perturbations of *Emiliana huxleyi* were studied under controlled conditions (Bellerby et al. 2008). Nutrient uptake showed no sensitivity to CO_2 treatment. There was no significant calcification response to changing CO_2 in *E. huxleyi* by the peak of the bloom and all treatments exhibited low particulate inorganic carbon production. Organic production increased with CO_2 production at the height of the bloom (Bellerby et al. 2008). Additional studies with *Emiliana huxleyi* showed that particulate organic production increased with rising CO_2 over a wide range of light intensities, but particulate inorganic production decreased with increasing CO_2 provided that light intensity was at least at a photon flux density of 150 $umol$ m^2/s, whereas below this light level it was unaffected by CO_2 (Zondervan et al. 2002). Under nutrient-sufficient conditions, an increase of CO_2 in seawater by twofold—a likely scenario by the end of the century—can significantly decrease both the rate of calcification by coccolithophorids and the ratio of inorganic to organic production (Sciandra et al. 2003). However, studies with *Emiliana huxleyi* grown under nitrogen limiting conditions—a situation that can also prevail in the ocean—showed a decrease in calcification rate, and in contrast to studies with N-replete cultures, gross community production and dark community respiration decreases. Increasing CO_2 has no measurable effect on the calcification/photosynthesis ratio when cells of *E. huxleyi* are NO_3-limited (Sciandra et al. 2003). The complex and often contradictory experimental results with coccolithophores in general, and *Emiliana huxleyi* in particular, requires identification of the dominant controls and responses to oceanic acidification (Bellerby et al. 2008).

5.2.3 Foraminifera

Two species of foraminiferans, *Orbulina universa* and *Globigerinoides sacculifer* subjected to 560 mg/1 of CO_2 experienced a 4 to 8% reduction in shell mass; at 780 mg/L this was 8 to 14% reduction in shell mass (as quoted in Fabry et al. 2008). Higher concentrations of

CO_2 in the atmosphere cause surface seawater to become more acidic and lower the calcium carbonate saturation rate. Foraminiferal shell weight for glacial-interglacial geological periods is directly related to seawater carbonate ion concentration, and supports the hypothesis that higher atmospheric carbon dioxide adversely affects marine calcification (Barker and Elderfield 2002).

5.2.4 Macroalgae

Not all photosynthetic species are adversely affected by increasing CO_2 enrichment (Gao et al. 1993; Kubler et al. 1999; Invers et al. 2002). Elevated CO_2 concentrations of 1,050 and 1,616 mg/L (adjusted to neutral pH) increased growth, photosynthesis, and nitrate uptake in two species of *Gracilaria* red alga, when compared to controls (345 mg CO_2/L); oxygen uptake was higher in the light than in the dark (Gao et al. 1993). Growth rate of the red seaweed, *Lomentaria articulata* increased with increasing CO_2 up to 200% of current ambient CO_2, but declined at 500% over ambient; growth was unaffected by ambient oxygen concentrations (Kubler et al. 1999). Plants collected in winter responded more extremely to CO_2 than did plants collected in the summer, although the overall pattern was the same. Biomass growth was limited by conversion of photosynthate to new biomass rather than by diffusion of CO_2, suggesting that non-bicarbonate-using macroalgae, such as *L. articulata*, may not be comparable to other higher plants in terms of their responses to changing gas composition (Kubler et al. 1999). The Mediterranean seagrass, *Posidonia oceanica*, could increase its depth limits as a consequence of increased CO_2 in the atmosphere and ocean (Invers et al. 2002). Studies showed that photosynthesis increased significantly at pH 7.5 over pH 8.2 with an 80% increase in net carbon balance at the lower pH. An increase in aqueous CO_2 will significantly enhance the carbon budget of *Posidonia* with subsequent colonization beyond its present depth limit (Invers et al. 2002).

Where water quality is not compromised elevated CO_2 derived from combustion of fossil fuel may increase the productivity and flowering rates of seagrasses worldwide (Palacios and Zimmerman 2007). Eelgrass, *Zostera marina*, and other seagrasses are expected to benefit from CO_2 enrichment (Bjork et al. 1997; Zimmerman et al. 1997; Palacios and Zimmerman 2007). Laboratory studies of one-year-duration with *Zostera* shoots growing under natural light-replete (33% surface irradiance) and light-limited (5% surface irradiance) were subjected to CO_2 concentrations equivalent to pH 8.1 (controls), 7.75, 7.7, or 6.2. Enrichment with CO_2 did not alter growth, leaf size, or leaf sugar content of above ground shoots in either light treatment. In light-replete conditions, however, CO_2 enrichment led to significantly higher reproductive output, below-ground biomass, and vegetative proliferation, suggesting that increasing the CO_2 content of the atmosphere and ocean surface will increase the productivity of seagrass meadows (Palacios and Zimmerman 2007). Seagrasses will probably flourish under acidic conditions (Zimmerman et al. 1997). Globally increasing CO_2 may enhance seagrass growth and survival in eutrophic coastal waters. Since seagrasses evolved within the last 90 million years when pH of ocean surface waters may have been as low as 7.4, it is suggested that high concentrations of free CO_2 would allow seagrasses to compete more effectively with macroalgae without the need for an efficient mechanism of bicarbonate utilization (Zimmerman et al. 1997). During the Cretaceous era, seagrasses had an affinity for inorganic carbon at least as high as macroalgae under the low pH and high CO_2/HCO_3^- concentration ratios, indicating that their photosynthetic capacity then matched that of macroalgae (Beer and Koch 1996). In the high pH and high CO_2/HCO_3^- ratios of today, their affinity for inorganic carbon is lower than that of macroalgae, suggesting that this deficiency inhibits use of inorganic carbon; this situation may be reversed again as global CO_2 levels in the atmosphere and the near-shore marine habitat increase in the future (Beer and Koch 1996).

The response of calcareous seagrass epibionts to elevated pCO_2 in aquaria and at a volcanic vent area, where seagrass habitat had been exposed to high CO_2 levels for decades, indicated that coralline algae—which were the dominant contributors to calcium carbonate mass on seagrass blades at normal pH—were absent from the system at mean pH 7.7, and were dissolved in aquaria enriched with CO_2 (Martin et al. 2008).

Decreasing pH was associated with altered transformation rates of toxic chemical species, increased metal uptake, and reduced growth rates. Dead biomass of the brown seaweed, *Ecklonia* sp., can reduce the toxic hexavalent chromium ion to the less toxic trivalent chromium ion, the rate of reduction being most rapid at pH <3.0 (Park et al. 2007). Decreased growth rate and increased copper content of *Thalassiosira pseudonana* were altered independently of medium copper concentration by decreasing pH (Sunda and Guillard 1976). Studies with *Ulva lactuca* show inhibited lead uptake with decreasing pH (Shiber and Washburn 1978); similarly, uptake of yttrium and strontium was enhanced as the pH of the medium decreased from 8.0 to 6.5 (Hampson 1967). Increasing accumulations of zinc in marine algae were also associated with decreasing pH (Eisler 1981). And salt marsh plants show enhanced mercury accumulations in roots at elevated pH (Gambrell et al. 1977).

5.2.5 Phytoplankton

Productivity of marine phytoplankton may double as a result of doubling of the atmospheric CO_2 concentration (Schippers et al. 2004). Authors predict a productivity increase up to 40% in those species with a low affinity for bicarbonate (Schippers et al. 2004). The supply of dissolved inorganic carbon—of which CO_2 comprises <1%—is not considered to limit oceanic primary productivity as its concentration in sea water exceeds that of other plant nutrients such as nitrate and phosphate. However, under controlled optimal light and nutrient conditions, diatom growth rate is limited by the supply of CO_2 (Riebesell et al. 1993). The doubling in surface water CO_2 levels since the last glaciation from 180 to 355 mg/L may have stimulated marine productivity, thereby increasing carbon sequestration (Riebesell et al. 1993). Mesocosm studies with natural phytoplankton communities show that dissolved inorganic carbon consumption increases with rising CO_2 (Riebesell et al. 2007). The community consumed up to 39% more dissolved inorganic carbon at three times present CO_2 levels, although nutrient uptake was the same. The carbon/nitrogen ratio increased from 6.0 at low CO_2 to 8.0 at high CO_2, exceeding the carbon/nitrogen ratio of 6.6 present in oceans today. This excess carbon consumption was associated with greater loss of organic carbon from upper layers of the stratified mesocosms, underscoring the importance and complexity of phytoplankton in ocean to global changes (Riebesell et al. 2007).

The effect of CO_2-induced seawater acidification on the concentration of inorganic nutrients (nitrate, nitrite, ammonium, soluble reactive phosphorus, and silicate) at a coastal site in the Western English Channel was studied between March and July 2008 (Wyatt et al. 2010). Ambient pH varied by 0.2 units over the study period. However, purging with CO_2 (380, 500, 760, and 1,000 mg/L) resulted in a maximum pH decrease of 0.4 units. Surface nitrate was depleted during the spring phytoplankton bloom and nitrogen limitation was prevalent thereafter. Acidification did not change the concentrations of nitrate, nitrite, soluble reactive phosphorus, or silicate; however, total ammonium increased 20% in acidified seawater during the pre-bloom period. A model based on these data shows that the region will become a net sink for ammonia if atmospheric CO_2 concentrations increase to 717 ppm, with a resultant increase in phytoplankton biomass (Wyatt et al. 2010).

Growth rate of a natural assemblage of mixed phytoplankton from Korea was measured at 250, 400, and 750 mg/L CO_2 in a series of mesocosms over 14 days (Kim et al. 2006). In all enclosures, two phytoplankton taxa (microflagellates and cryptomonads) and two species of diatoms (*Skeletonema costatum*, and *Nitzschia* spp.) comprised about 90% of the phytoplankton community. During the nutrient-replete period from days 9 to 14, both diatom populations increased substantially; however, only *Skeletonema costatum* showed an increasing growth rate with increasing seawater CO_2 (Kim et al. 2006). Phenotypic consequences of selection for growth at elevated CO_2 concentrations was investigated for the unicellular green alga *Chlamydomonas reinhardtii* (Collins and Bell 2004). After about 1,000 generations, selection lines of the alga failed to evolve specific adaptation to a CO_2 concentration of 1,050 mg/L. Some lines, however, evolved showing high rates of photosynthesis and respiration combined with higher chlorophyll content and reduced cell size. These lines also grew poorly at ambient (430 mg/L) concentrations of CO_2, possibly due to the accumulation of mutations in genes affecting the carbon concentration mechanism. In short, 1,000 generations of selection at high CO_2 concentrations produced no increase in growth at high CO_2, whereas growth at ambient CO_2 was often markedly reduced (Collins and Bell 2004). Variable concentrations of dissolved molecular carbon dioxide affects C:N:P ratios in six of seven species of phytoplankton tested, and correlated with a CO_2-dependent decrease in growth rate

(Burkhardt et al. 1999). No general pattern for CO_2-related changes in elemental composition were found with regard to the direction of trends, and are unlikely to have a significant effect on the oceanic carbon cycle (Burkhardt et al. 1999).

Phytoplankton include the Cyanobacteria. Two species of cyanobacteria, *Synechococcus* sp. and *Prochlorococcus* sp., were incubated under present day (380 ppm) and predicted year-2100 CO_2 levels (750 ppm) and under normal versus elevated (+4°C) temperatures (Fu et al. 2007). Increased temperature stimulated cell division rates of *Synechococcus* but not *Prochlorococcus*. Doubled CO_2 combined with elevated temperature increased maximum photosynthetic rates of *Synechococcus* four times over controls. Temperature also altered other photosynthetic parameters in *Synechococcus* but these changes were not observed in *Prochlorococcus*. Authors suggest that global change would influence the dominance of *Synechococcus* and *Prochlorococcus* ecotypes, with likely effects on oligotrophic food webs. Authors state that individual cyanobacteria strains may respond quite differently to future CO_2 and temperature increases and that caution is needed when generalizing their responses to global changes in the ocean (Fu et al. 2007). Some photosynthetic organisms appear to benefit from increasing CO_2, including *Trichodesmium*, a nitrogen-fixing cyanobacterium, which responded with increased carbon and nitrogen fixation at elevated CO_2 (Czerny et al. 2009). The mechanism underlying this CO_2 stimulation is unknown. Contrary results were obtained with *Nodularia spumigena*, a heterocystous bloom-forming nitrogen fixing cyanobacterium from the Baltic Sea. *Nodularia* reacted to hypercapnia with reduced cell division rates and nitrogen fixation rates, and altered carbon, phosphorus, and elemental composition of the formed biomass (Czerny et al. 2009).

Responses of phycocyanin-rich and phycoerythrin-rich strains of *Synechococcus* differed on exposure to CO_2 concentrations of 350, 680, or 800 mg/L for 12 days (Lu et al. 2006). For example, growth of the phycoerythrin-rich strain was unaffected by CO_2 enrichment, whereas the phycocyanin-rich strain grown at 800 mg/L CO_2 experienced a 36.7% increase in growth over the batch grown at 350 mg/L. On the other hand, the phycocyanin-rich strain showed no change in carbohydrate content over the CO_2 range measured, whereas the phycoerythrin-rich strain exhibited a CO_2-induced increase of 37.4% at 800 mg/L. Moreover, both strains showed increases in cellular pigment contents in CO_2-enriched treatments than controls (Lu et al. 2006).

Physiological responses of iron-replete and iron-limited cultures of the cyanobacterium *Crocosphaera* was evaluated at 190 (glacial), 380 (current), or 750 (projected year 2100) mg/L CO_2 (Fu et al. 2008). Rates of nitrogen and carbon dioxide fixation and growth increased with increasing CO_2, but only under iron-replete conditions. Nitrogen and carbon fixation rates at 750 mg/L were 1.4–1.8-fold and 1.2–2.0-fold, respectively, relative to those at present day and glacial pCO_2 levels. In iron-replete cultures, cellular iron and molybdenum quotas varied threefold and were linearly related to nitrogen fixation rates and to external pCO_2 levels. Higher CO_2 and iron concentrations both resulted in increased cellular pigment contents and affected photosynthesis-irradiance interactions. If these results apply to natural *Crocosphaera* populations, anthropogenic CO_2 enrichment could increase global oceanic nitrogen and carbon dioxide fixation, depending on iron availability. Possible biogeochemical consequences may include elevated inputs of nitrogen to the ocean and increased potential for iron and phosphorus limitation in the future high CO_2 ocean, and feedbacks to atmospheric pCO_2 in the near future and over glacial to interglacial time scales (Fu et al. 2008).

5.3 Invertebrates

Ocean acidification has been proposed to pose a major threat to marine biota, particularly shell-forming and calcifying organisms (Hendriks et al. 2010). However, analysis of organism responses to elevated CO_2 suggests that marine biota may be more resistant to ocean acidification than expected. Active biological processes and small-scale temporal and spatial variability in ocean pH may render calcifiers far more resistant to ocean acidification than previously expected (Hendriks et al. 2010).

Effects of CO_2-acidified seawater on structure and diversity of adults from macrofaunal and nematode assemblages, and of sediment nutrient fluxes, was investigated under different regimens of pH (8.0, 7.3, 6.5, 5.6), exposure (2 or 20 weeks), or sediment types (muddy vs. sandy) in mesocosms (Widdicombe et al. 2009). Exposure to acidified seawater significantly altered community structure and reduced diversity for both macrofaunal and nematode assemblages. Impacts on sandy sediment fauna were greater than those on muddy sediment fauna. Sandy sediments also showed the greatest effects on nutrient fluxes with increasing acidification, with efflux of nitrate, nitrite, and silicate decreased and efflux of ammonium increased. In mud, acidification increased the efflux of ammonium but had no effect on other nutrients. Leakage from carbon storage facilities and ocean acidification could cause significant changes in coastal sediment communities and that lowered seawater pH could also affect nutrient cycling directly by altering bacterial communities (Widdicombe et al. 2009).

5.3.1 Protists

Introduction of CO_2 arising from flue discharges of fossil fuel-fired power plants directly into the ocean with pH plumes as low as 6.0 may cause inhibition of marine nitrification (Huesemann et al. 2002). The rate of microbial ammonia oxidation was reduced by about 50% at pH 7 when compared to pH 8.0 and more than 90% at pH 6.5, with accumulation of ammonia instead of nitrate (Huesemann et al. 2002).

5.3.2 Coelenterates

Changes to the stability of coastal reefs may reduce the protection they offer to coastal communities against storm surges and hurricanes (USDC 2008). Coastal and marine commercial fishing in United States waters generates as much as 30 billion dollars per year and nearly 70,000 jobs (USDC 2008). Healthy coral reefs are the foundation of many of these viable fisheries, as well as the source of tourism and recreation revenues. Nearly half the U.S.-managed fisheries depend on coral reefs and related habitats for a portion of their life cycles yielding an estimated value to U.S. fish stocks of 250 million dollars annually (USDC 2008). Weakening of coral reef structures occurs from a changing ocean chemistry wherein the oceans absorb part of the excess atmospheric CO_2 (Lough 2008). Coral reefs will not disappear but their appearance, structure and community composition will radically change. Greenhouse gas mitigation strategies are needed to prevent the full consequences of human activities causing these alterations to coral reef ecosystems (Lough 2008).

Coral reefs were one of the first ecosystems to be recognized as vulnerable to ocean acidification (Kleypas and Yates 2009). Compared to pre-industrial rates, warm-water corals with symbiotic algae are predicted to calcify 10 to 50% less by the year 2050. The decline in calcium carbonate production, coupled with an increase in calcium carbonate dissolution will diminish reef building and the benefits that reefs provide, such as breakwaters for shoreline protection, and habitat for mangroves and seagrass beds (Kleypas and Yates 2009). Global degradation of coral reef ecosystems has resulted in heavy loss of adult corals (Hoegh-Guldberg 1999; Gardner et al. 2003; Hughes et al. 2003). The persistence and recovery of coral reefs require that recruitment keeps pace with loss of adults; unfortunately surveys indicate low levels of recruitment throughout the Florida Keys and the Caribbean (Porter and Meier 1992; Hughes and Tanner 2000; Gardner et al. 2003; Shearer and Coffroth 2006). Decreased calcification of corals is also associated with increasing oceanic absorption of CO_2 (Kleypas et al. 1999). About 20% of the coral reefs have been destroyed in the past few decades and another 50% are verging on collapse (Stone 2007). One of the chief causes of decline is the increased ocean acidity as rising carbonic acid levels deplete carbonate. By mid-century atmospheric CO_2 levels could reach more than 500 ppm and near the end of the century 800 ppm. The latter figure would decrease surface water pH by about 0.4 units, lowering the pH to a level near 20 million years ago. Some coral species may assume an anemone-like form to survive, but by pH 7.9 most reefs would be gone (as quoted in Stone 2007). Under natural, undisturbed, nutrient-limited conditions, elevated CO_2 depresses calcification, stimulates the rate of turnover of organic carbon, especially in the light, but has no effect on net organic production (Langdon et al. 2003). Some scientists, however, disagree with these findings. Adey (1998), for example, states that warm-water coral reefs are significant sinks of ocean alkalinity and will probably not be affected by fossil carbon introduction into the earth's atmosphere during the next few centuries. Bicarbonate additions dramatically stimulate photosynthesis and calcification in selected corals (Herfort et al. 2008). In the next century, the predicted increase in CO_2 will result in about a 15% increase in oceanic HCO_3^-, and this could stimulate photosynthesis and calcification in a wide variety of reef-building corals. The current contradictory reports about the likely effects of CO_2 addition and oceanic acidification on corals and other calcifiers indicate that predictions need to be reexamined (Herfort et al. 2008).

Scleractinian cold-water corals—corals that lack symbiotic algae—form, perhaps, the most vulnerable marine ecosystems to hypercapnia (Turley et al. 2007). They are found throughout the world oceans, usually between 200 and 1,000+ m in depth, live for several centuries, form reef frameworks that persist for millennia, and are thought to experience relatively little environmental variability. The decline in surface ocean pH with declines in carbonate ion concentration, aragonite, and calcite, will seriously affect marine calcifiers, especially aragonitic organisms such as scleractinian corals. The scleractinian corals have survived several mass extinction events, but in all cases took several millions of years to recover. Perturbations in the carbon cycle, most likely resulting in ocean acidification, played a fundamental role in all major mass extinctions of this group. The extremely rapid release of anthropogenic CO_2 from fossil fuel deposits is unprecedented in geological history and risks perturbing deep-water coral ecosystems before the scientific community has begun to map and understand them (Turley et al. 2007). The gradual decline in growth rate of juvenile scleractinian corals from a reef in St. John, U.S. Virgin Islands, has been documented over a 10-year period and is attributed to a depressed aragonite saturation state and rising seawater temperatures (Edmunds 2007). Calcification rates of the cold-water coral *Lophelia pertusa* were affected by ambient pH (Maier et al. 2009). Among colonies of *Lophelia pertusa*, calcification rates were highest in the youngest polyps with up to 1% daily (mean 0.11% daily) in new skeletal growth. Lowering the pH by 0.15 and 0.3 units relative to ambient pH resulted in a strong decrease in calcification by 30 and 56%, respectively. The effect of changes in pH reduction was strongest for faster-growing young polyps (59% reduction) than for older polyps (40% reduction), suggesting that skeletal growth of the youngest and fastest calcifying corallites are most sensitive to ocean acidification. Nevertheless, all colonies of *L. pertusa* studied showed a positive net calcification regardless of ambient pH lowering of 0.3 units, indicating some adaptation to a low pH environment (Maier et al. 2009).

Fertilization, settlement, and growth rates of the threatened Caribbean coral, *Acropora palmata*, were all negatively impacted by increasing CO_2 and impairment of fertilization was exacerbated at lower sperm concentrations (Albright et al. 2010). The cumulative impact of ocean acidification on fertilization and settlement success is an estimated 52% and 73% reduction in the number of larval settlers on the reef under CO_2 conditions projected for the middle (560 mg/L CO_2) and the end of this century (800 mg/L CO_2), respectively. Additional declines of 39% (mid-CO_2) and 50% (high CO_2) were observed in post-settlement rates when compared to controls. These results suggest that ocean acidification has the potential to impact multiple, sequential early life history stages, severely compromising sexual recruitment and the ability of coral reefs to recover from disturbance (Albright et al. 2010). Intracellular pH of incubated cells of the reef coral *Stylophora pistillata* and the symbiotic anemone *Anemonia viridis* were measured in short-term light and dark-incubated cells (Venn et al. 2009). In all cells isolated from both species, intracellular pH was markedly lower than the surrounding seawater pH of 8.1. In cells that contained symbiotic algae, mean intracellular pH values were significantly higher in light-treated cells than dark treated cells: 7.41 vs 7.13 for the coral and 7.29 vs.7.01 for the anemone. In contrast there was no difference in intracellular pH in light and dark treated cells without algal symbionts (Venn et al. 2009).

Carbon dioxide and temperature significantly interact on physiology of the scleractinian coral *Stylophora pistillata*, with calcification, as one example, reduced 50% at elevated temperatures and elevated CO_2 concentrations; this interaction could explain a large portion of the variability documented for coral physiological variables and CO_2 concentrations (Reynaud et al. 2003). In estimates of coral reef calcification based on alkalinity anomalies, interference from changes in nitrate, ammonium, and sulfate ions is likely to cause errors as high as 15% (Kinsey 1978). Acidification is expected to reduce coral reef calcification and increase reef dissolution. Inorganic cementation—a process wherein $CaCO_3$ occludes porosity and binds framework components—is also affected, especially in the eastern tropical Pacific (Manzello et al. 2008). This region of naturally low pH waters (7.88–7.98) has a low carbonate saturation rate and trace abundances of cement. Low cement abundances may be a factor in the bioerosion rate of these reefs (Manzello et al. 2008).

Coral reef mesocosms subjected to increasing CO_2 to effect a pH drop from 8.1 to 7.8 for up to 30 days showed a decreasing calcification rate and a decreasing aragonite saturation rate (Leclercq et al. 2002). A 10-month controlled experiment was conducted to test the impact of ambient concentrations of carbon dioxide (382 mg/L) and elevated (747 mg/L) levels on common calcifying reef organisms from Kaneohe Bay, Hawaii (Jokiel et al. 2008). Acidification adversely affected development of crustose coralline algae populations by 86% and their growth by 250%. Coral calcification decreased between 15% and 20% under acidified conditions. However, larvae of the coral *Pocillopora damicornis* were able to recruit under the acidified conditions, and gamete production in the coral *Montipora capitata* was unaffected by acidification after 6 months of treatment (Jokiel et al. 2008).

Model results based on a conservative scenario of atmospheric CO_2 increase were used to examine changes in sea surface temperature and aragonite saturation state over the Pacific Ocean basin until 2069 when CO_2 levels are predicted to approach 517 ppm (Guinotte et al. 2003). Rising atmospheric CO_2 concentrations will reduce the saturation state of carbonate minerals in the surface ocean over the next 70 years until nearly all locations become marginal with respect to calcification, reef accumulation, and changes in community structure. However, the effect of high temperature and low saturation state is unknown at present; this combination of stressors has probably not occurred since the Pleistocene (Guinotte et al. 2002). Some corals might survive large-scale environmental change, such as that expected during the year 2100 (Fine and Tchernov 2007). This conclusion was based on a 12-month laboratory study during which scleractinian coral colonies of *Oculina patagonica* (encrusting) and *Madracis pharencis* (bulbous) were subjected to pH values of either 7.3 to 7.6 or 8.0 to 8.3 (ambient). After 1 month of acidic conditions, polyps became elongated followed by complete skeleton dissolution, although polyps remained attached to the undissolved hard rocky substrate. The biomass of the solitary polyps was three times as high as biomass of control polyps which continued to calcify and grow. Controls and 90% of the acidic groups maintained their algal symbionts, but all seemed normal after two months. Gametogenesis in both groups developed similarly during spring and summer months. All skeleton-free coral fragments survived for the entire 12-month period. After 12 months, when transferred back to ambient pH conditions, the acid-challenged soft-bodied polyps calcified and reformed colonies (Fine and Tchernov 2007). Fine and Tchernov's decalcification experiments may not be representative of all varieties of corals (Stanley 2007). Zooxanthelate reef-building species would have responded differently to the experiment because of the complex nature of their photosymbiosis (Stanley 2007).

Laboratory studies with fragments of living *Porites lutea* demonstrated that calcification rate in daylight showed a linear increase with increasing aragonite saturation rate (Ω) values of seawater (Ohde and Hossain 2004). At high aragonite saturation rates, corals also calcified during darkness. Authors conclude that calcification of *Porites lutea* depends on Ω of seawater. A decrease in saturation state of seawater due to increased pCO_2 may decrease reef-building capacity of corals through a reduction in calcification rate (Ohde and Hossain 2004). Slow-growing cold water corals are especially sensitive to acidification. Reefs are composed of aragonite, a carbonate material that is more soluble than the calcite used by coccolithophores and sea urchins. Carbonate, in turn, is more soluble than calcite in high pressures and cold water. By the end of the century, two-thirds of deep-water corals—as opposed to virtually none today—would be exposed to sea water that is corrosive to aragonite (as quoted in Ruttimann 2006). Polyps of the common Atlantic golf ball coral *Favia fragum* were reared for 8 days at 25°C in seawater with aragonite saturation states ranging from ambient (Ω = 3.7; pH 8.2) to strongly undersaturated (Ω = 0.22; pH 7.6) (Cohen et al. 2009). Aragonite was taken up by all corals, even those reared in strongly undersaturated seawater. However, low aragonite/pH groups, when compared to controls, had delayed initiation of calcification and subsequent growth of the primary corallite, and altered composition of the aragonite crystals (13% increase in Sr/Ca, and 45% decrease in Mg/Ca). Observed changes in crystal morphology are consistent with a >80% decrease in the amount of aragonite precipitated by corals. This suggests that the saturation state of fluid within the isolated calcifying compartment, while maintained by the corals at levels above that of the external seawater, decreased systematically and significantly as the saturation state of the external seawater decreased . Future impact of ocean acidification on tropical coral ecosystems depends on the ability of individuals or species to achieve the levels of calcifying fluid supersaturation required to ensure rapid growth (Cohen et al. 2009).

Nubbins of the hermatypic coral *Porites compressa* grown over 5 weeks at pH 7.2 calcified at half the rate of control corals grown at pH 8.0 (Marubini and Atkinson 1999). Corals in low pH treatment recovered their initial calcification rates within 2 days of re-introduction to ambient seawater, suggesting that the effects of CO_2 chemistry are immediate and reversible. Changes in calcification from increases in atmospheric CO_2, and hence decreases in CO_3^{2-}, may be larger than effects from elevated nutrients (Marubini and Atkinson 1999). Decreasing carbonate-ion concentration suppresses skeletal growth in scleractinian corals (Marubini et al. 2002). In one 8-day study, four physiologically different species of of hermatypic corals (*Acropora verweyi, Galaxea fascicularis, Pavona cactus,* and *Turbinaria reniformis*) were cultured under normal (280 umol/kg) and low (140 umol/kg) carbonate-ion concentrations. The low carbonate treatment resulted in a significant suppression of 13–18% of calcification rate for all species and a tendency for weaker crystallization at the distal tips of fibers, which was most pronounced in *A. verweyi,* and least in *T. reniformis* (Marubini et al. 2002). Ocean acidification will compromise carbonate accretion, with corals becoming increasingly rare on reef systems (Hoegh-Guldberg et al. 2007). Serious consequences are predicted for reef-associated fisheries, tourism, coastal protection, and people (Hoegh-Guldberg et al. 2007). Nevertheless, the relative roles of climate and anthropogenic activities on reef declines should be reexamined (Buddemeier and Ware 2003).

Under laboratory conditions, staghorn corals, *Acropora intermedia,* and massive corals, *Porites lobata.* were subjected to either 388 mg CO_2/L (controls, pH 8.0–8.4), 520–700 mg/L (intermediate, pH 7.85–7.95), or 1,000–1,300 mg CO_2/L (high, pH 7.60–7.70) at 25–28°C or 28–29°C (Anthony et al. 2008). After 8 weeks, the high CO_2 *Acropora* groups had 40–50% loss of pigmentation (bleaching); for *Porites* it was 20%. Pigmentation loss for the intermediate *Acropora* group was 20%. Increasing CO_2 usually led to reductions in daily productivity, as measured by photosynthesis less respiration, and a significant reduction in calcification when compared to controls. In all cases, effects were exacerbated at the elevated temperatures. Anthony et al. (2008) conclude that elevated concentrations of CO_2 are bleaching agents for corals under high irradiance acting synergistically with warming to lower thermal bleaching thresholds, and that acidification had a greater effect on bleaching and productivity than on calcification.

5.3.3 Molluscs

Among bivalve molluscs adverse effects of declining pH on various species are documented (Calabrese and Davis 1966; Kuwatani and Nishii 1969; Bamber 1987, 1990; Ringwood and Keppler 2002; Michaelidis et al. 2005; Berge et al. 2006; Lannig et al. 2010; Zippay and Hofmann 2010). Calabrese and Davis (1966) found that the pH range for normal embryonic development of the American oyster, *Crassostrea virginica,* was 6.75 to 8.75 and for quahog clams, *Mercenaria mercenaria,* it was 7.00 to 8.75; the lower pH limit for survival of oyster larvae was 6.00 and for clam larvae 6.25. The growth rate of larvae of both species was inhibited at pH levels below 6.75, with optimal growth recorded for clam larvae at 7.5–8.0 and for oyster larvae at 8.25–8.5 (Calabrese and Davis 1966). Based on 7-day field studies, growth rates of juvenile quahog clams were reduced by more than 50% when average pH fell below 7.5 or minimum pH levels fell below 7.2 (Ringwood and Keppler 2002). Bamber (1987) exposed young carpet-shell clams *Venerupis decussata* for up to 30 days to seawater ranging between 3.5 and 8.2 (controls). Shell dissolution occurred at pH <7.55; feeding was inhibited at pH <7.0 and both tissue and shell growth were inhibited at this pH ; more than 50% died at pH <6.5, with smaller clams (3–4 mm) more sensitive than larger (7–9 mm) clams. Author concluded that minor changes in pH below those found normally in the sea are intolerable to young *Venerupis decussata* (Bamber 1987). Studies with the European oysters, *Ostrea edulis,* Pacific oysters, *Crassostrea gigas,* and the common mussel, *Mytilus edulis,* held for up to 30 days at various pH levels showed reduced survival and growth, and abnormal behavior at comparatively low pH (Bamber 1990). Many deaths occurred at pH <6.0 among Pacific oysters, mussels at pH <6.6, and among European oysters at pH <6.9 after 60 days. Survival was reduced at a given pH level with increasing time of exposure, increasing temperature, and increasing size of the animal. At pH <7.0, growth suppression, tissue weight loss, reduced shell size, shell dissolution, or suppressed feeding activity were observed in all species. Abnormal behavior (shell gaping, torpor) occurred at pH <6.5. Author concludes that seawater of pH <7.9 is intolerable to bivalve molluscs (Bamber 1990). Unlike arthropods and echinoderms, sperm motility of Pacific oysters was relatively unaffected by pH; highest motility was obtained in the range of pH 4.0 to 12.0 (Dong et al. 2002). Japanese pearl oysters, *Pinctada fucata*, of age one year were reared for 40 days at pH regimens of 7.36, 7.48, 7.66, 7.78, 7.89, 8.07, and 8.04 (controls) (Kuwatani and Nishii 1969). Growth was best at pH 8.07 and 8.04. Mortality was heavy at pH 7.36 and 7.48, with most deaths occurring before 20 days. Among the remaining groups decreasing pH was associated with decreasing growth and increased shell dissolution (Kuwatani and Nishii 1969). Pacific oysters held at pH 7.7 for one month, when compared to controls at elevated pH had lowered hemolymph and increased CO_2 levels, decreased mantle alanine and ATP, and increased gill succinate (Lannig et al. 2010). On acute warming from 15°, metabolism increased in both groups. Altered metabolic pathways following acidification-reduced pH exposure suggests that increasing temperature associated with climate change may affect oysters and other populations of sessile coastal invertebrates (Lannig et al. 2010).

Synergistic effects of elevated CO_2 concentrations and water temperature on fertilization and embryonic development of the Sydney rock oyster *Saccostrea glomerata* were documented under controlled conditions (Parker et al. 2009). As CO_2 increased, fertilization significantly decreased, along with decreased veliger growth and a higher percentage of abnormal larvae. As temperature increased above 26°C or decreased below 26°C, embryonic development was inhibited. The proportion of abnormal veligers was greatest at 1,000 mg/L CO_2 and 18–30°C (>90%) and least at 375 mg/L CO_2 and 26°C (<4%). There was no development at 30°C and 750–1,000 mg/L CO_2, suggesting that predicted changes in ocean acidification and temperature over the next century will adversely affect reproduction and development of oysters and other marine invertebrates (Parker et al. 2009). However, near-future levels of ocean acidification (pH 7.80) alone did not affect sperm motility and fertilization kinetics in the Pacific oyster, *Crassostrea gigas* (Havenhand and Schlegel 2009). Veliger larvae of the eastern oyster, *Crassostrea virginica*, and the suminoe oyster, *Crassostrea ariakensis* were grown in estuarine waters for 28 days under four CO_2

regimes: 280, 380, 560, and 800 mg/L (Miller et al. 2009). These CO_2 regimes simulated atmospheric conditions in the pre-industrial era, present, and projected future concentrations in 50 and 100 years, respectively. Eastern oysters experienced a 16% decrease in shell area and a 42% reduction in calcium content when pre-industrial (280 mg/L) and end of 21st century (800 mg/L) were compared. Suminoe oysters showed no change in growth or calcification. Both species demonstrated net calcification and growth, even when aragonite was undersaturated, a result that runs counter to previous expectations for invertebrate larvae that produce aragonite shells (Miller et al. 2009). The pH of some Chesapeake Bay tributaries that once supported large populations of oysters, *Crassostrea virginica*, is increasing (Waldbusser et al. 2010). However, the pH of some tributaries now correspond to values found in laboratory studies that reduce oyster biocalcification rates or resulted in net shell dissolution. Biocalcification declined significantly with a reduction of 0.5 pH units and higher (Waldbusser et al. 2010).

Mussels, *Mytilus edulis*, were exposed to CO_2-acidified seawater (pH 7.8, 7.6, or 6.5, control = pH 8.0) for 60 days (Beesley et al. 2008). Seawater acidification significantly reduced mussel health, as measured by the neutral red retention assay for lysosomal membrane stability, possibly owing to elevated levels of calcium ions in the hemolymph generated by the dissolution of the mussels' calcium carbonate shells. Other tissue structures were unaffected. However, predicted long-term changes to seawater chemistry associated with ocean acidification are likely to have a more significant effect on health and survival of *M. edulis* populations than the short-lived effects envisaged from CO_2 uptake leakage from sub-seabed storage (Beesley et al. 2008). Mussels, *Mytilus galloprovincialis*, were subjected to seawater of pH 7.3 (reduced from 8.05 through the addition of CO_2) for 90 days (Michaelidis et al. 2005). Long-term hypercapnia caused a permanent reduction in hemolymph pH. Acidosis was limited through increased hemolymph bicarbonate levels derived from the dissolution of shell $CaCO_3$. Oxygen consumption fell, metabolism was reduced, growth was inhibited, nitrogen excretion increased, and protein degradation was enhanced. A reduction in seawater pH to 7.3 may be fatal to mussels and a seawater pH above 7.5 is recommended for all shelled molluscs (Michaelidis et al. 2005). Studies with the common mussel *Mytilus edulis* subjected for 44 days to various CO_2-induced pH regimens (6.7 to 8.1) showed no growth at pH 6.7, reduced growth at pH 7.1, and normal growth at 7.4–8.1 (Berge et al. 2006). Effects of hypercapnia for 32 days on the immune response of *Mytilus edulis* were observed in mussels exposed to to CO_2-acidified sea water of pH 7.8 (controls), 7.7, 7.5 or 6.7 (Bibby et al. 2008). Levels of phagocytosis increased significantly during the exposure period, but was suppressed on acidification. Acidified seawater had no measurable effect on other parameters measured including superoxide anion production, and total and differential cell counts. Hemocytes are affected by ocean acidification and may impact calcium pathways (Bibby et al. 2008). Calcification rate of the mussel, *Mytilus edulis*, declined 25% at 740 mg/L CO_2 (projected CO_2 concentration for the year 2100) and total dissolution projected at 1,800 mg/L CO_2 (Gazeau et al. 2007). Calcification rate of Pacific oysters declined linearly with increasing CO_2, reaching 10% at 740 mg/L CO_2 (Gazeau et al. 2007).

Current and future increases in pelagic CO_2 concentration may deplete or alter the composition of shellfish populations in coastal areas, based on studies with larvae of clams *Mercenaria mercenaria*, bay scallops *Argopecten irradians*, and eastern oysters *Crassostrea virginica* held at 650 mg/L CO_2. Clams and scallops had low survival (<50%), delayed metamorphosis, and smaller sizes. All oysters survived, but experienced lowered growth and delayed metamorphosis; oyster survival, however, was diminished at 1,500 mg/L CO_2 and higher. (Talmage and Gobler 2009). Larvae of *Mercenaria mercenaria* and *Argopecten irradians* were grown at 250, 390, or 1,500 mg/L CO_2 and effects on growth, survival, and condition were monitored (Talmage and Gobler 2010). Larvae grown at 250 mg/L CO_2 (a pre-industrial level) displayed significantly faster growth and metamorphosis as well as higher survival and lipid accumulation rates when compared to modern day CO_2 levels, *viz.*, 390 mg/L. Bivalves grown at 250 mg/L displayed thicker, more robust shells than those grown at 390 mg/L, whereas bivalves exposed to CO_2 levels expected later this century, *viz.*, 1,500 mg/L had shells that were malformed and eroded. The ocean acidification that has occurred during the past two centuries may inhibit development and survival of larval shellfish and could contribute to global declines of some bivalve populations (Talmage and Gobler 2010).

Fertilized eggs of the Pacific oyster, *Crassostrea gigas* were incubated for 48 h in CO_2-acidified seawater of pH 7.4—the projected oceanic pH in the year 2300—and compared to a control group maintained at pH 8.1 (Kurihara et al. 2007). Veligers from the CO_2 group had significantly lower survival, growth, shell mineralization, and increased abnormal development when compared to controls (Kurihara et al. 2007). In Willapa Bay, Washington, spat of Pacific oysters had unusually high mortality between 2005 and 2008 (Welch 2009). Deaths were linked to a decline in pH of offshore waters that upwell near shore, causing dissolution of oyster shells and a favorable environment for the pathogen *Vibrio tubiashii* (Welch 2009). Ecological dynamics may confound the overall picture Declining pH will negatively affect calcareous species, including a reduction in density of predatory and grazing species of molluscs, thus affording a degree of protection to noncalcareous fleshy algae and certain pH-resistant arthropods food items (Wootton et al. 2008). At a pH of 7.4 and lower, bivalve molluscs experience significant growth reduction rates, shell dissolution, reduced metabolism, and increased mortality; at 740 mg CO_2/L there is a 10%–25% decrease in calcification rate (as quoted in Fabry et al. 2008).

Among gastropod molluscs, the intertidal *Littorina littorea* responded to low seawater pH by a depression in metabolic rate, production of a thinner shell—thus making them more vulnerable to predation—and an increase in avoidance behavior (Bibby et al. 2007). Biological effects from ocean acidification may be complex and extend beyond simple direct effects (Bibby et al. 2007). In *Littorina obtusata* encapsulated embryos, seawater of pH 7.6 produced reduced viability, increased developmental times, lower heart rates, and altered shell morphology (Ellis et al. 2009). Pteropod gastropods, *Limacina helicina*—a key Arctic pelagic mollusc—were kept in culture at CO_2 levels of 350 mg/L (pH 8.09; present value) or 760 mg/L (pH 7.78; pH expected for the year 2100; Comeau et al. 2009). Shell aragonite calcification showed a 28% decrease at the pH value expected for the year 2100 compared to the present pH value, suggesting that a population decline would likely cause dramatic changes in carbon and carbonate cycling in polar ecosystems (Comeau et al. 2009). Shallow water marine gastropods subjected to a 200 mg/L increase in CO_2 for 6 months (total CO_2 of about 550 mg/L) had reduced growth (Shirayama and Thornton 2005). Juveniles of two species of gastropod abalones were subjected to various pH regimens from 6.79 to 9.01 for 57 to 68 days (Harris et al. 1999). For the greenlip abalone, *Haliotis laevigata*, maximum growth rate was recorded at pH 8.27, with some growth reduction outside the range 7.78–8.77. For the blacklip abalone, *Haliotis rubra*, growth was best in the range 7.93 to 8.46 with some growth reduction outside this range. For both species, survival was significantly reduced at pH 6.79. Some deaths were also recorded at pH 9.01 and pH 7.16 (Harris et al. 1999). Larvae of the red abalone, *Haliotis rufescens*, held in CO_2-acidified seawater (pH 7.87) show decreased thermal tolerance of pretorsion and late veliger stages when compared to controls (pH 8.05), but other developmental stages were unaffected (Zippay and Hofmann 2010). Decreased pH had no measurable effect on the expression pattern of the two shell formation genes in any of the abalone larval stages. The differential sensitivity of larval stages to low pH needs to be considered in long-term resiliency of individual species to environmental changes (Zippay and Hofmann 2010).

Decapod molluscs are comparatively resistant to CO_2 and cuttlefish, *Sepia officinalis*, are unusually resistant to elevated CO_2 and low pH (Gutowska et al. 2008). During a 6-week period, juvenile cuttlefish maintained calcification under 4,000 mg/L (pH 7.23) and 6,000 mg/L CO_2 (pH 7.10) and grew at the same rate as controls (628–705 mg/L CO_2, pH 7.94–8.01). The mechanistic processes are imperfectly understood (Gutowska et al. 2008). Additional studies with cuttlefish embryos (Gutowska and Melzner 2009) show that oxygen, CO_2, and pH values in perivitelline fluid all show a linear relationship with increasing wet mass: oxygen declined, CO_2 increased, and pH decreased from 7.7 to 7.2. Oxygen is limiting in cephalopod embryos towards the end of development but measured CO_2 and pH values in cuttlefish eggs would be harmful to other species of marine ectothermic animals (Gutowska and Melzner 2009). Acute CO_2 tolerance of juveniles of the kisslip cuttlefish *Sepia lycidas*, and oval squid *Sepioteuthis lessoniana* are also comparatively elevated: 50% survival at 8,400 mg/L CO_2 in 24 h for cuttlefish, and for squid 5,900 mg/L in 24 h and 3,800 mg/L CO_2 in 48 h (Kikkawa et al. 2008).

Results of acidification on molluscan survival and metal uptake are uneven. Some studies with lead and molybdenum show that decreasing pH increases uptake and adversely affects survival; other studies with silver and zinc show different trends. Studies with lead and American oysters, *Crassostrea virginica*, showed that up to 90% of lead added to seawater was detected in the particulate fraction and the rest as a complex in the soluble fraction (Zaroogian et al.

1979). The adsorptive behavior of lead is strongly pH dependent. Thus, when the pH is lowered, lead is released from the particulates and shift occurs in equilibriums between chemical species of lead. This change in equilibrium favors ionic lead in the oyster digestive tract which has a pH range of 5.5 to 6.0 (Zaroogian et al. 1979). Studies with the clam, *Venerupis pallustra,* and molybdenum show that decreasing pH of medium affects molybdenum residues and survival, with higher residues and lower survival reported in the pH range 5.1–6.2 (Abbott 1977). The picture is different for silver and zinc. Silver associated with food of estuarine organisms was unavailable for incorporation due to silver's ability to adsorb rapidly to cell surfaces and to remain tightly bound despite changes in pH (Connell et al. 1991). Duke (1967) reports that pH of the ambient medium significantly affects radiozinc uptake from seawater by a community of oysters, clams, and scallops; however, uptake was unexpectedly higher in all cases at elevated pH levels.

5.3.4 Nemertea

Short-term laboratory studies of 96-hour duration with the nemertean worm *Procephalothrix simulus,* demonstrated that all survived a pH range of 5.00–9.20; however, deaths were recorded at pH 4.70 and lower and at 9.50 and higher (Zhao and Sun 2006).

5.3.5 Annelids

Polychaete worms, *Nereis virens,* were subjected for 5 weeks to seawater acidified from pH 7.9 (controls) to a pH of 7.3, 6.5, or 5.6 with CO_2 gas (Widdicombe and Needham 2007). These treatments mimicked the effects of either ocean acidification (pH 7.3) or leakage from a seabed CO_2 storage site (pH 6.5 and 5.6). The size and structure of worm burrows were unaffected by the acidity, but acidity had a profound effect on sediment nutrient flux including nitrate, nitrite, phosphate, and ammonium ion. Ocean acidification significantly affects sediment nutrient flux in coastal and shelf seas as a result of changes in structure and function of bioturbating communities (Widdicombe and Needham 2007). In another study, *Nereis virens* were subjected to a pH range of 5.1 to 8.1 (controls) for 10 days at 18°C (Batten and Bamber 1996). Deleterious effects at pH<6.5 were evident as shown by increased mortality, decreased burrowing activity, and reduced growth; survivors below pH 6.0 had significantly reduced glycogen levels. When the study was repeated at 9°C for 30 days, the worms were comparatively inactive with high survival, but with low body weight below pH 6.5. Glycogen levels gave the clearest indication of stress, with a marked decline below pH 6.5. Low pH levels are deleterious to *N. virens* and body glycogen levels are useful where whole animal responses may be poorly indicative (Batten and Bamber 1996).

A sipunculid worm, *Sipunculus nudus,* subjected to 10,000 mg/L CO_2 experienced metabolic suppression and high mortality during 7-week exposure (Portner et al. 1998). However, longer exposures of 67 days to lower CO_2 concentrations also produced death attributed to disturbances in homeostatic regulation (Langenbuch and Portner 2004). Studies with *Sipunculus nudus* showed that a decrease in extracellular pH from 7.9 to 6.7 caused a reduction of 40%–45% in aerobic metabolic rate, an increased accumulation of intracellular bicarbonate, a reduction in amino acid metabolism, and a shift in the selection of amino acids used favoring monoamino dicarboxylic acids and their amines, specifically asparagine, glutamine, aspartic acid, and glutamic acid (Langenbuch and Portner 2002).

5.3.6 Arthropods

Acid-iron wastes discharged into the New York Bight have negligible effect on copepod survival owing to the very few minutes in which lethal concentrations of low pH persist (Grice et al. 1973). The iron floc which persists in the acid grounds at great dilutions does not affect adult copepods or their developmental stages. Nevertheless, zooplankton biomass of the acid grounds was 30% lower than that of a control area for unknown reasons (Grice et al. 1973). Acute toxicity of lowered pH to six species of copepods (*Calanus pacificus, Neocalanus cristatus, Eucalanus bungii bungii, Pseudocalanus minutus, Metridia pacifica, Paraeuchaeta elongata*) an ostracod (*Conchoecia* sp.), and a euphausid (*Euphausia pacifica*) was determined (Yamada and Ikeda 1999). Sensitivity was species specific with 50% mortality after 96 hours in the pH range of 5.8–6.6 and no deaths in the range 6.6–7.8. Lowered pH had no measurable effect on swimming behavior, diet, and size of gills (Yamada and Ikeda 1999).

Eggs of the copepod *Acartia tsuensis* were exposed to 2,380 mg/L CO_2 through maturity and over two subsequent generations (Kurihara and Ishimatsu 2008). Compared to controls (380 mg/L CO_2), exposure through all life stages of first generation copepods had the same survival, body size, and rate of development. Egg production and hatching rates were also not significantly different between the initial generation of females exposed to high CO_2 and 1st and 2nd generation females developed from eggs to maturity in high CO_2. Copepods appear more tolerant to CO_2 than sea urchins and bivalve molluscs. However, the importance of copepods in marine ecosystems requires thorough evaluation of the overall environmental changes predicted to occur with increased CO_2 concentrations including increased temperature, enhanced UV irradiation and changes in the community structure and nutritional value of phytoplankton (Kurihara and Ishimatsu 2008). Mortality of copepods increased with increasing CO_2 and increasing exposure time in species collected from both shallow (surface) and deep layers (1,500 m) in subarctic, transitional, and subtropical regions (Watanabe et al. 2006). Deep-living copepods showed higher tolerance to CO_2 than did shallow-living copepods. Moreover, deep-living copepods from subarctic and transitional regions had higher tolerances to CO_2 than did subtropical copepods. The higher tolerances of the deep-living copepods from subarctic and transitional regions are attributed to their adaptation to the comparatively elevated natural CO_2 conditions in the subarctic ocean (Watanabe et al. 2006).

Egg production and biomass loss in adult female copepods, *Calanus finmarchicus*, was not affected by simulated ocean acidification of pH 6.95 and 8,000 ppm CO_2 representative of the worst case atmospheric CO_2 scenario for the year 2300 (Mayor et al. 2007). But only 4% of the eggs successfully yielded nauplii larvae after 72 hours in the experimental treatment—significantly lower than the controls which were maintained at pH 8.23—demonstrating that reproduction in *Calanus finmarchicus*, and other calanoid copepods is pH sensitive (Mayor et al. 2007). Two species of copepods (*Acartia* spp.) subjected to 365 mg CO_2/L (controls, pH 8.1), 2,000 mg CO_2/L (pH 7.3), 5,000 mg CO_2/L (pH 7.0) or 10,000 mg CO_2/L (pH 6.8) exhibited decreased egg hatching rate and increasing nauplius mortality rate with increasing CO_2 concentration (Kurihara et al. 2004, 2004a). Liquid carbon dioxide was introduced into the sea floor at a depth of 3,250 m and sampled 2 and 40 m from the deposition site after 30 days (Thistle et al. 2005). The pore water pH of the deposition site was 0.75 unit lower (more acidic) than in samples taken further away. Numbers of harpacticoid copepods and other species of representative deep-sea fauna were reduced at the deposition site when compared with conspecifics from distant sites (Thistle et al. 2005). Gut pH of the copepod *Calanus helgolandicus*, was determined under a range of feeding conditions (Pond et al. 1995). Median gut pH of the fore- and hind- guts of starved individuals was 6.86 and 7.19, respectively. Copepods feeding on diatoms, dinoflagellates, or coccolithophorids (all of which had a median gut pH > 7.0) had foregut pH as low as 6.11, suggesting that the foregut is the site of acid secretion (Pond et al. 1995). Calcium carbonate dissolution in copepod gut is governed by pH and is highest when individual copepods alternate between grazing and non-grazing and feeding is restricted to the night time period (Jansen and Wolf-Gladrow 2001). Up to 70% of the ingested carbonate is dissolved in gut and 15% dissolution is expected in a bloom situation. The most critical parameter in this model is gut pH, as only pH values <6.5 lead to significant dissolution (Jansen and Wolf-Gladrow 2001). Grazing and fecal pellet production by the copepods *Calanus helgolandicus* and *Pseudocalanus elongatus* feeding on the coccolithophore *Emiliania huxleyi* were studied in the laboratory (Harris 1994). The vertical flux of inorganic carbon was considered significant. Author found that 27% to 50% of the ingested coccolith calcite was egested in fecal pellets, and these sank at an average rate of 100 m daily (Harris 1994). Decreasing pH results in decreasing survival and larval development rate of copepods challenged with copper (Sunda et al. 1990).

Effects of CO_2-acidified seawater on embryonic development of the intertidal amphipod *Echinogammarus marinus* included a more protracted embryonic development at low pH, although the effect was only evident at low salinity (Egilsdottir et al. 2009). However, reduced salinity, not pH, exerted a strong significant effect on numbers and calcium content of hatchlings. Ocean acidification may affect aspects of amphipod development but exposure to realistic salinities appear, in the short term, to be more important in impacting development than exposure to CO_2-acidified seawater at levels predicted 300 years hence (Egilsdottir et al. 2009). Deep-sea amphipods, *Eurythenes obesus*, from hydrothermal vents were especially abundant near the Loihi seamount near Hawaii where diluted vent waters had a mean pH of 6.3 (minimum 5.7) and a temperature of 5°C (Vetter and Smith 2005). Amphipods exposed for 60 minutes to diluted vent waters became very active within seconds and then were narcotized over the next 12–15 minutes; however, all amphipods revived within 30 minutes of removal from the plume (Vetter and Smith 2005). The amphipod *Gammarus locusta*, which has a cosmopolitan distribution in estuaries, is comparatively tolerant to pH of 7.8 (550 mg/L CO_2) and pH 7.6 (980 mg/L CO_2) when compared to pH 8.1 (about 380 mg/L CO_2) (Hauton et al. 2009). *Gammarus locusta* juveniles were reared to maturity (about 28 days) in seawater of pH 7.6, 7.8, and 8.1 with no differences between groups in growth, survival, and expression of genes for a heat shock protein (*hsp70* gene). However, there was a consistent and significant increase in expression of the *gapdh* gene (which governs enzyme metabolism of glyceraldehyde-3-phosphate dehydrogenase) at pH 7.5, suggesting that metabolic changes may occur in response to acidification and that subtle effects on physiology and metabolism of coastal marine species may be overlooked in studies of organism growth and mortality (Hauton et al. 2009).

Chronic exposure to elevated CO_2 seawater (pH 7.4) of the intertidal barnacle *Amphibalanus amphitrite* affects some, but not all, aspects of discrete life stages (McDonald et al. 2009). There were no effects of reduced pH on larval condition, cyprid size, cyprid attachment and metamorphosis, juvenile to adult growth, or egg production when compared to controls reared at pH 8.2. Barnacles reared at pH 7.4 had larger basal shell diameters suggestive of compensatory calcification, but their central wall plates required less force to penetrate than controls. Thus, dissolution rapidly weakens wall shells as they grow, making them more vulnerable to predation (McDonald et al. 2009). Effects of CO_2-induced acidification on survival, shell mineralogy, embryonic development and the timing of larval release were investigated in the intertidal barnacle *Semibalanus balanoides* (Findlay et al. 2009). Compared to controls (CO_2 = 344 mg/L, pH = 8.07) adult survival was 22% lower in the high-CO_2 treatment (CO_2 = 922 mg/L, pH =7.70) with significant changes in the mineral structure of the adult shell. Embryonic development rate was slower in the high-CO_2 treatment than in the control but within rates found in natural populations from similar locations. There was an estimated 19-day delay in development under high-CO_2 conditions, which resulted in a 60% reduction in the number of nauplii reaching hatching stage when over 50% of the control nauplii had hatched. At 922 mg/L CO_2 some adults were able to survive and larvae were able to hatch, indicating that there is still potential for this barnacle to find suitable habitats and for populations to develop and survive (Findlay et al. 2009). Growth and development of metamorphosing post-larvae of *Semibalanus balanoides*, were measured at the northern edge of their geographic distribution in the Arctic Ocean where temperatures were increasing and pH decreasing (Findlay et al. 2010). Growth and development were negatively impacted at lower pH (7.7) compared to controls (pH 8.1), but were not affected by elevated temperatures (+4°C). The mineral composition of the shells was the same regardless of treatment. The combination of reduced growth and maintained mineral content suggests a change in the energetic balance of the exposed animals. The idea of reallocation of resources under different conditions of pH merits further investigation (Findlay et al. 2010). Post-larval development of intertidal barnacles is affected by elevated CO_2, elevated temperature, and inherent species differences (Findlay et al. 2010a). Shell mineralogy and survival of post-larvae of *Semibalanus balanoides* and *Elminius modestus* were measured in intertidal microcosms at 380 mg/L CO_2 (pH 8.07) or 1,000 mg/L CO_2 (pH 7.70), 14 or 19°C, over a period of 30 days. Control growth rates, using operculum diameter, were 14 um daily for *S. balanoides* and 6 um daily for *E. modestus*. Growth rate decreased in *E modestus* in high CO_2, but shell calcium content and survival were unaffected by either elevated temperature or CO_2. Growth of *S. balanoides* larvae was variable, but shell calcium content and survival was reduced under elevated temperature and CO_2. A decrease of 0.4 pH units alone would not be sufficient to impact survival of barnacles during the first month post-settlement, but in conjunction with a 4–5°C increase in temperature significant changes to the biology of these organisms will occur (Findlay et al. 2010a).

A euphausid, *Euphausia pacifica*, subjected to pH 7.6 and lower, had increasing mortality with increasing exposure time and decreasing pH; a similar pattern was documented for copepods subjected to 860 mg/L CO_2 (as quoted in Fabry et al. 2008). In the bathypelagic mysid, *Gnathophausia ingens*, pH in the range of 7.1 to 8.7 did not affect oxygen consumption rate or activity (Mickel and Childress 1978). The percent oxygen extraction was significantly greater at pH 7.1 than at pH 7.9 or 8.7, the organism was unable to regulate oxygen consumption in the pH range studied. Authors conclude that since the increase in percent oxygen extraction at pH 7.1 does not improve the ability of *G. ingens* to regulate its oxygen uptake, it would appear that there is a loss in effectiveness elsewhere in its respiratory system at this pH (Mickel and Childress 1978). Embryos and larvae of the Antarctic krill, *Euphausia superba*, were subjected to 380 (control), 1,000 or 2,000 mg/L CO_2 with no apparent effect on development or behavior at 1,000 mg/L (Kawaguchi et al. 2010). However, at 2,000 mg/L, embryonic development was disrupted in 90% before gastrulation with zero hatch. Since CO_2 in the southern Ocean is predicted to reach 1,400 mg/L by the year 2100, additional research is merited on krill developmental and later stages in order to predict the possible fate of this key species in the Southern Ocean (Kawaguchi et al. 2010).

Juveniles of the migratory school prawn *Metapenaeus macleayi* begin to avoid acidified seawater at pH 5.9 (Kroon 2005). This catadromous species migrates to the ocean to spawn, with migratory pathways blocked by acid sulfate soil discharges of pH 5.0 (Kroon 2005). Predicted future seawater CO_2 conditions have been demonstrated to adversely affect survival, growth, and reproduction of the shrimp *Palaemon pacificus*, and possibly other crustacean populations (Kurihara et al. 2008). In their study, shrimps were reared in seawater containing 1,000 mg/L of CO_2 (pH 7.89) for 30 weeks. Survival, egg production, growth, and molting frequency were reduced at 1,900 mg/L (pH 7.64) after exposure for 15 weeks (Kurihara et al. 2008). Prawns, *Penaeus occidentalis* and *Penaeus monodon* were exposed for 36–56 days in flowing seawater enriched with gaseous carbon dioxide at pH levels ranging from 6.4 to 7.9 (Wickins 1984). Exposure to hypercapnic seawater reduced growth and decreased frequency of molting. Calcium levels in cuticle increased with increasing exposure but strontium and magnesium were unchanged (Wickins 1984). Juvenile *Penaeus monodon* were comparatively resistant to low pH, with a 96-hour LC50 of pH 3.7 and with growth reduced 5% over a 23-day period at pH 5.9 (Allan and Maguire 1992). Prawns held at pH 5.5 for 23 days had significantly decreased dry matter, reduced growth, and increased molting frequency when compared to those held at pH 7.8 (Allan and Maguire 1992). Deep sea crustaceans are also comparatively resistant to CO_2. Juveniles of the kuruma prawn, *Marsupenaeus japonicus*, have 50% survival on exposure to 14,300 mg/L CO_2 for

72 h, suggesting an inverse relationship between oxygen requirement and CO_2 tolerance among marine fauna (Kikkawa et al. 2008). In tidal salt marshes of South Carolina, low oxygen and high CO_2 conditions occur frequently; however, respiratory responses of the grass shrimp, *Palaemonetes pugio*, were unaffected over a wide range of environmental fluctuations in CO_2 and O_2 (Cochran and Burnett 1996). Elevated CO_2 and low oxygen decreases the resistance of Pacific white shrimp, *Litopenaeus vannamei*, to a pathogenic strain of the bacterium *Vibrio campbelli* (Burgents et al. 2005). Among shrimp injected with a sublethal dose of *V. campbelli*, both hypercapnia and hypoxia (low oxygen) increased the distribution of the bacterium to the hepatopancreas and gills—major targets for the pathogenic effects of *Vibrio* spp. (Burgents et al. 2005). Reduced pH is associated with a reduction in arthropod survival, inhibited larval development, reduced sperm motility, and increased metal uptake. Copper uptake by brine shrimp, *Artemia franciscana*, increases with decreasing pH (Blust et al. 1991).

The shallow water Dungeness crab, *Cancer magister*, when subjected to a high dose of carbon dioxide recovered more quickly than the deep sea Tanner crab *Chionocetes bairdi* similarly exposed, suggesting that deep-sea animals are more sensitive to acidification changes than shallow water animals (Ruttimann 2006). Based on studies with Dungeness and Tanner crabs, deep-sea animals have reduced metabolic rates and lack the short-term acid-base regulatory capacity to cope with acute hypercapnic stress that would accompany large scale CO_2 sequestration (Pane and Barry 2007). Additionally, data indicate that CO_2 sequestration in oxygen-poor areas of the ocean would be even more detrimental to deep-sea fauna (Pane and Barry 2007). Hypercapnia causes enhanced sensitivity to heat and thus a narrowing of the thermal tolerance window in the edible crab, *Cancer pagurus* (Metzger et al. 2007). Short-term studies with crabs demonstrated that exposure to 1,000 mg/L CO_2 (water pH dropped from 7.9 to 7.06) caused a significant reduction of oxygen partial pressure in the hemolymph, as well as a large 5°C downward shift of upper thermal limits. Interactions of ambient temperature and anthropogenic increases in ambient CO_2 concentrations should be considered in future studies of climate change effects to ecosystems (Metzger et al. 2007). The effect of elevated CO_2 concentrations (380, 710, or 3,000 mg/L CO_2) on thermal tolerance of the spider crab *Hyas araneus* demonstrated that the critical temperature was

>25°C at 380 mg/L CO_2, 23.5°C at 710 mg/L CO_2, and 21.1°C at 3,000 mg/L CO_2 (Walther et al. 2009). CO_2-induced acidification has the potential to cause a narrowing of thermal windows; a further increase in ambient temperature, combined with ocean acidification may cause *Hyas* to reach their physiological limits even sooner (Walther et al. 2009). The effect of different levels of hypercapnia-induced (i.e., CO_2-induced) acidification on extracellular acid base balance of the velvet swimming crab, *Necora puber*, was investigated over a period of 16 days (Spicer et al. 2007). In the pH range tested (8.0, 7.3, 6.7, 6.1) any extracellular acidosis incurred was compensated by an increase in bicarbonate supplied mostly by dissolution of the exoskeleton. After 16 days, however, crabs subjected to the lowest pH regimes exhibited some extracelluar acidosis. This species appears comparatively resistant to low pH. But local acidification as a result of ocean CO_2 dispersal or leakage from geological sequestration is likely to compromise even this species (Spicer et al. 2007). Blue crabs, *Callinectes sapidus*, exposed to low oxygen and low pH have suppressed phenoloxidase activity, making them more susceptible to infectious pathogens (Tanner et al. 2006). Phenoloxidase activity decreased with pH, showing a 16% reduction at pH 7.0 from a normoxic pH 7.8. Authors suggest that decreased phenoloxidase activity at low tissue O_2 and pH compromises the ability of crustaceans in hypercapnic hypoxia to defend themselves against microbial pathogens (Tanner et al. 2006). Molybdenum is relatively toxic to the crabs *Eupagurus bernhardus* and *Carcinus maenas* at alkaline pH typical of seawater, but death—and presumably accumulation—occur at pH values near 5.0 (Abbott 1977).

Sperm motility of the horseshoe crab, *Limulus polyphemus*, is affected by reduced pH caused by zinc deficiency (Clapper et al. 1985 b). Larvae of the European lobster, *Homarus gammarus* were cultured in CO_2-acidified seawater containing approximately 1,200 mg/L CO_2 at pH 8.10 vs. controls at 315 mg CO_2/L and pH 8.39 (Arnold et al. 2009). Survival, carapace length, and zoeal progression were all normal. However, carapace mass was reduced, as was calcium and magnesium content of the carapace. Alterations are most likely the result of acidosis or hypercapnia interfering with the normal homeostatic function, and not a direct impact on the carbonate supply-side of calcification. These alterations could adversely affect the competitive fitness and recruitment success of of larval lobsters (Arnold et al. 2009).

5.3.7 Echinoderms

Ocean acidification as a result of increased atmospheric CO_2 is predicted to lower the pH of seawater to between pH 7.6 and 7.8 over the next 100 years, with the greatest changes expected in polar waters (Clark et al. 2009). Effects of lowered pH on sea urchin pluteus larvae were examined under controlled conditions. Effects of lowered pH on larvae from tropical *(Tripneustes gratilla)*, temperate *(Pseudechinus huttoni, Evechinus chloroticus)*, and a polar species *(Sterechinus neumayeri)* were examined. Larvae were reared in seawater of pH 6.0, 6.5, 7.0, 7.5, 7.7, 7.8 and 8.1–8.2. The tropical larvae were tested at 26°C for 4 days; the temperate species at 10–15°C for 9–13 days ; and the polar species at –1°C for 7 days. Lowering pH to less than 7.0 resulted in: increased mortality for all species; reduced size of larvae, although external morphology was unaffected; and a significant reduction in calcification of the larval skeleton of 14–37%, with the exception of the polar species which showed no significant difference (Clark et al. 2009). High mortality and test dissolution was recorded at pH 7.8 and lower for various species of sea urchins; 500 mg/L CO_2, and higher, decreased urchin fertilization rates and larval development (as quoted in Fabry et al. 2008).

When developing sea urchin embryos were reared under different pH regimens, larval differentiation was sharply affected by a moderate pH decrease of 0.5 units (from pH 8.1) (Pagano et al. 1985). Even pH decreases as small as 0.2 units showed early and irreversible damage to embryogenesis. Moreover, mitotic abnormalities were observed following early exposure to decreased pH. However, increased pH up to 8.6 failed to exert any adverse effect on development (Pagano et al. 1985). Purple-tipped sea urchins, *Psammechinus miliaris*, were exposed to artificially acidified seawater treatments of pH 6.16, 6.63, 7.44 or 8.1 (controls) for 8 days (Miles et al. 2007). All died at pH 6.16 of acidosis after 7 days. Coelomic fluid of survivors from other groups when compared to controls, showed an accumulation of CO_2, a significant reduction in pH, and increases in magnesium suggestive of test dissolution.

Bicarbonate buffering was used to reduce the acidosis, but compensation was incomplete. A chronic reduction of surface water pH to <7.5 would be detrimental to this and other intertidal species of echinoids (Miles et al. 2007). In another study, larvae of the sea urchin, *Lytechinus pictus*, were raised from fertilization to the pluteus stage in seawater with elevated CO_2 with emphasis on skeletogenesis and gene expression (O'Donnell et al. 2010). During this 150-hour period, larvae were exposed to either 380 mg/L CO_2 and pH 7.93 = control; 540 mg/L CO_2 and pH 7.87; or 970 mg/L CO_2 and pH 7.78. Those grown in a high CO_2 environment were smaller and had a more triangular body shape than those raised in normal CO_2 conditions. Gene expression profiling showed that genes central to energy metabolism and biomineralization were down-regulated in the larvae in response to elevated CO_2, whereas only a few genes involved in ion regulation and acid-base balance were affected. Results suggest that although larvae are able to form an endoskeleton, development at elevated CO_2 levels has consequences for larval physiology as shown by changes in the larval transcriptome (O'Donnell et al. 2010).

Adults of the green sea urchin, *Strongylocentrotus droebachiensis*, exposed for 8 weeks to high sublethal concentrations of CO_2, when compared to a control group, had significantly reduced gonad growth (by 67%), and significantly reduced feed intake and total food consumption (Siikavuoplo et al. 2007). The reduction in gonad growth was attributed to a decrease in feed intake and impaired feed conversion efficiency. Adult green sea urchins show low tolerance to increased carbon dioxide levels and are unable to maintain high gonad growth under these conditions (Siikavuoplo et al. 2007).

Sperm agglutination in the sea urchin, *Arbacia punctata*, was most frequent at pH levels below 7.5 and above 8.5 (Gregg and Metz 1976). Increasing CO_2 concentrations on the early development of the sea urchins *Hemicentrotus pulcherrimus* and *Echinometra mathaei* were investigated (Kurihara et al. 2004, 2004a; Kurihara and Shirayama 2004). Fertilization rate, cleavage rate, developmental speed and pluteus larval size all tended to decrease with increasing CO_2 concentration beginning at 1,000 mg/L of CO_2 equivalent to a pH of 7.61; at 10,000 mg/L CO_2 (pH 6.8), the fertilization rate was reduced 50% over controls. Decreased pH and increased seawater CO_2 concentrations adversely affect marine ecosystems (Kurihara et al. 2004, 2004a; Kurihara and Shirayama 2004). Shallow water sea urchins and other species held in 200 mg/L CO_2 above ambient (total of about 550 mg/L CO_2) for 6 months had reduced growth (Shirayama and Thornton 2005).

The acid-base status of two intertidal sea urchins *Psammechinus miliaris* and *Echinus esculentus* was studied during periods in air when removed from the ambient medium (Spicer et al. 1988). The carbon dioxide capacity of the coelomic fluid of both was low and only marginally greater than that of sea water. The pH of the coelomic fluid was also low (7.05–7.17) and was influenced mainly by the internal pressure of CO_2. Acid-base disturbance in the coelomic fluid of both species during emersion was minimal. The coelomic fluid of both species was in a state of perfectly compensated respiratory acidosis; however, an increase in the concentration of Ca^{2+} and Mg^{2+} ions may be related to the dissolution of test as a source of carbonate buffer (Spicer et al. 1988). Upon exposure to air (emersion) the purple sea urchin *Strongylocentrotus purpuratus* releases fluid from the esophagus causing air to appear within the test, occupying 33% of the volume of the intrathecal space (Burnett et al. 2002). The intestine containing air forms a facultative lung and contributes to the oxygenation of the perivisceral coelomic fluid during emersion. Authors suggest that compensation for respiratory acidosis induced by air exposure does not occur in organisms that are unable to regulate ions in a dilute environment and that the facultative lung ensures a minimal partial pressure of oxygen in perivisceral coelomic fluid (Burnett et al. 2002).

Sperm flagellar motility in broadcast spawning reef invertebrates—including sea cucumbers (echinoderm) and corals (coelenterate)—was seriously impaired below pH 7.7, suggesting that fertilization taking place in seawater may decline in the foreseeable future (Morita et al. 2010).

Low pH-induced inhibition of early development and fertilization were observed in various species of echinoderms. Gametes and larvae of the sea urchin *Heliocidaris erythrogramma* were exposed to CO_2-induced acidification wherein the pH was lowered by 0.4 pH units—the upper limit of predictions for the year 2100—with significant reductions in sperm motility, sperm swimming speed, and fertilization success (Havenhand et al. 2008). Larval survival and development of the brittlestar *Ophiothrix fragilis* was adversely affected by a CO_2-driven acidification drop in pH of 0.2 units from 8.1 to 7.9 (Dupont et al. 2008). Larvae of the purple sea urchin, *Strongylocentrotus purpuratus*, were subjected to 380 mg/L CO_2 (pH 8.01 = control), 540 mg/L CO_2 (pH 7.96), or 1,020 mg/L CO_2 (pH 7.88) from fertilization through 70- hours post-fertilization (Todgham and Hofmann 2009). At elevated CO_2 concentrations, larvae showed significantly lower mRNA transcript levels of genes central to biomineralization, the cellular stress response, metabolism, and programmed cell death. Physiological processes beyond calcification are impacted greatly suggesting a need for additional genomics-based studies (Todgham and Hofmann 2009).

In the case of the ophiuroid brittlestar *Amphiura filiformis* (some with one or two arms excised) exposed for 40 days to pH 8.0 (controls), pH 7.7 (a 2099 scenario), pH 7.3 (a projected 2300 estimate), or pH 6.8, acid-stressed brittlestars showed increased metabolism and increased calcification, potentially ameliorating effects of increased acidity (Wood et al. 2008). However, there was increasing muscle loss with increasing acidity suggesting that this process is unlikely to be sustainable in the long term (Wood et al. 2008). In another 40-day study with *Amphiura filiformis* under the same four pH regimens, their role in nutrient cycling (through bioturbation of sediments) is affected, especially nitrate fluxes (Wood et al. 2009). At pH 6.8, silicate and phosphate fluxes increased, and muscle loss continued to increase (Wood et al. 2009).

Early research on effects of near- future ocean acidification on echinoderm larvae generally show negative effects, such as decreased growth rate, increased mortality, and developmental abnormalities. However, all long-term studies were performed on feeding planktotrophic larvae (single parent produces millions of small eggs that develop into planktotrophic larvae feeding on exogenous sources, such as phytoplankton or dissolved organic matter) while alternative life-history strategies, such as nonfeeding lecithotrophy (parents produce only 2,000–6,000 eggs with large yolk reserves) were largely ignored (Dupont et al. 2010a). For example, lecithotrophic larvae and juveniles of the common sun star, *Crossaster papposus*, are positively impacted by ocean acidification. When cultured at low pH (7.7, equivalent to 980 mg/L CO_2 vs. controls at pH 8.1 or about 390 mg/L CO_2) larvae and resultant juveniles grow faster with no effect on survival or skeletogenesis. This suggests that in future oceans, lecithotrophic species may be better adapted to acidified oceans than planktotrophic species, with important consequences at the ecosystem level (Dupont et al. 2010a).

Nevertheless, near-future ocean acidification is expected to have negative impact on echinoderm taxa with significant consequences at the ecosystem level (Dupont et al. 2010b).

The interactions between pH, CO_2 , and thermal stress in echinoderms are unusually complex (Gooding et al. 2009; O'Donnell et al. 2009). In one case, the sea star *Pisaster ochraceus* showed increasing growth and feeding rates with increasing temperature from 5 to 21°C and with increasing CO_2 from 380 to 780 mg/L (Gooding et al. 2009). On the other hand, larvae of the sea urchin *Strongylocentrotus franciscanus* raised at 540 and 970 mg/L CO_2 then subjected to 1-h acute temperature stress up to 31°C were unable to mount a physiological response (as measured by expression of the gene *hsp70*) relative to those raised under ambient CO_2 conditions (O'Donnell et al. 2009). Zinc and molybdenum are known to modify acidification effects (Abbott 1977; Clapper et al. 1985a). Acidification adversely affects survival and sperm motility of selected echinoderms; however, reduced sperm pH and motility in sea urchins was associated with zinc deficiency (Clapper et al. 1985a). The starfish, *Asterias rubens*, is relatively resistant to molybdenum salts, with 50% dead reported in 24 hours at 127.0–254.0 mg Mo/L; however, mortality was highest at the lower end of the pH range tested of 5.5 to 8.2 (Abbott 1977).

To investigate whether the presence of a burrowing urchin, *Echinocardium cordatum*, might influence the effect of ocean acidification on subtidal sediment pH profiles and nematode community structure , an experiment was conducted using subtidal sediments with or without urchins and seawater at either pH 8.0 (ambient) or 7.5 (Dashfield et al. 2008). The presence of urchins and a reduction in pH both had significant effects on within-sediment pH profile. When urchins were present at pH 8.0, sediment pH was lower than that of overlying seawater and nematode abundance was higher. It was concluded that ocean acidification can lead to changes in nematode communities in subtidal sediments affected by burrowing urchins (Dashfield et al. 2008).

5.3.8 Bryozoans

In the field, bryozoans were the only calcifiers present on seagrass blades at mean pH 7.7 where the total mass of epiphytic calcium carbonate was 90% lower than that of pH 8.2 (Martin et al. 2008).

5.3.9 Chaetognaths

Chaetognaths, *Sagitta elegans*, held at pH 7.6 and lower showed increasing mortality with increasing exposure time and decreasing pH (as quoted in Fabry et al. 2008). Mortality of *S. elegans* was 50% after 72 h at pH 6.73 and no deaths at pH 7.76 (Omori et al. 1998).

5.4 Vertebrates

This group include the sharks, rays, bony fishes, marine reptiles, birds, and mammals.

5.4.1 Elasmobranchs and Bony Fishes

Fishery resources contribute15% of animal protein for three billion people worldwide and a further one billion people rely on fisheries for their primary source of protein (UNEP 2010). Fish stocks, already declining in many areas due to overfishing and habitat destruction, now face the new threats posed by ocean acidification. The effect of increased acidity on adult finfish—but not earlier developmental stages—seems minimal in the species investigated. However, their orientation and balance mechanisms as well as behavior may be sensitive to ocean acidification (UNEP 2010). Fish are more susceptible to a rise in environmental CO_2 than terrestrial animals because the difference in CO_2 partial pressure of the body fluid of fish and of the ambient medium is smaller by orders of magnitude than in terrestrial animals (Ishimatsu et al. 2005). Hypercapnia adversely affects respiration, circulation, and metabolism. Changes in these functions are likely to reduce growth rate and population size through reproduction failure, and change the distribution pattern due to avoidance of high-CO_2 waters or reduced swimming activities (Ishimatsu et al. 2005).

The dogfish, *Scyliorhinus canicula*, experienced increased ventilation rates at pH 7.7 and died within 72 h after exposure to 70,000 mg/L CO_2 (as quoted in Fabry et al. 2008). A pH range of 7.25–7.3 adversely affected diet intake of *Sparus aurata* and *Dicentrarchus labrax* , two species of bony fishes. Rapid death of adults, eggs, and larvae occurred at 50,000–150,000 mg/L CO_2 of Japanese whiting *Sillago japonica*, olive flounder *Paralichthys olivaceus*, eastern little tuna *Euthynnus affinis* , red sea bream *Pagrus major*, and amberjack

Seriola quinqueradiata (as quoted in Fabry et al. 2008). Juvenile sea bass *Dicentrarchus labrax* subjected for only 3 days to seawater acidified to 8.03 with CO_2 produces a compensatory alkalosis; increased plasma HCO_3—attributable to a net gain of bicarbonate ions from the medium—leads to increased blood pH (Cecchini et al. 2001). As expected, there was a negative correlation between feed intake and CO_2, and this could result in reduced growth and poor condition (Cecchini et al. 2001).

Marine fish when compared with invertebrate groups such as molluscs and corals are likely to be less affected by an increase in oceanic CO_2 or a decrease in pH; however, results of laboratory studies demonstrate that decreasing pH and increasing CO_2 dramatically influence physiology, metabolism, and reproductive biology of fish, with egg fertilization and survival of early developmental stages most vulnerable (Rijnsdorp et al. 2009). Acute CO_2 tolerance during early developmental stages of four marine species differed between species and stages (Kikkawa et al. 2003). Further, seawater acidified either by hydrochloric acid or by CO_2 to a constant pH may underestimate the toxic effects of CO_2 (Kikkawa et al. 2004). In that study, eggs and larvae of the sea bream, *Pagrus major*, were subjected to pH 8.1 (controls), pH 6.2 (produced by a 130-fold increase in CO_2), or pH 5.9 (a 260-fold increase of CO_2) for 6 hours (eggs) or 24 h (larvae). Egg mortality was 4% at pH 6.2 in the hydrochloric acid group and 86% in the CO_2 group; at pH 5.9, these values were 1% and 97%, respectively. For larvae, mortality was 2% at pH 6.2 in the HCl group and 61% in the CO_2 group; at pH 5.9, these values were 5% and 100%, respectively (Kikkawa et al. 2004). Twenty species of marine teleosts were able to discriminate a pH reduction of 0.04–0.06 units through the addition of CO_2 (Bull 1940). Using CO_2 to lower the pH of seawater, survival of larval plaice, *Pleuronectes platessa*, the most sensitive species tested, was nil below pH 6.15–6.5 (Bishai 1960). Rainbow trout survival is unaffected in the seawater range of pH 5.5 to 9.0; however, trout that survived pH 4.9 went blind (Carter 1964). Early studies with eggs and larvae of herring, *Clupea harengus*, conducted at 16.5o/oo and 8.0°C, showed that decreasing pH was associated with decreasing fertilization success, decreasing egg survival, decreasing growth rate, increasing heart rate, increasing larval abnormalities , and impaired prey-capture ability (Kinne and Rosenthal 1967). In short-term studies juveniles and adults of sand smelt *Atherina boyeri* avoided sea water of pH 6.5–6.6. but not 7.0 and higher (Davies 1991). Acidification impairs olfactory discrimination and homing ability of larval orange clownfish *Amphiprion percula* (Munday et al. 2009). Larval clownfish reared in control seawater (pH 8.15) discriminated between a range of olfactory cues that enable them to locate suitable reef habitat. At pH 7.8 (1,000 mg/L CO_2), larvae were attracted to olfactory stimuli they normally avoided and at pH 7.6 (1,700 mg/L CO_2) they no longer responded to any olfactory cues. At all pH levels tested, larvae had the same morphological appearance, swam and fed normally, had similar settlement behavior, and showed no evidence of histopathology (Munday et al. 2009).

Eggs and larvae of the orange clownfish were reared in seawater simulating a range of ocean acidification scenarios over the next 50–100 years: 390 mg/L CO_2, pH 8.06 (current); 538 mg/L CO_2, pH 7.94; 744 mg/L CO_2, pH 7.88; and 1,024 mg/L CO_2, pH 7.84 (Munday et al. 2009a). CO_2 acidification had no detectable effect on embryonic duration, and egg survival or size at hatch. Larvae demonstrated increased growth rate in acidified waters. Elevated CO_2 and reduced pH had no effect on the maximum swimming speed of settlement-stage larvae. Levels of ocean acidification likely to be experienced in the near future significantly disadvantage the growth and performance of benthic-spawning marine teleosts (Munday et al. 2009a). Levels of dissolved CO_2 predicted to occur in the ocean this century may alter the behavior of larval clownfish and decrease survival during recruitment to adult populations (Munday et al. 2010). Altered behavior was detected at 700 mg/L CO_2 with many individuals becoming attracted to the smell of predators. At 850 mg/L CO_2 the ability to sense predators was completely impaired. Larvae exposed to elevated CO_2 were more active and exhibited riskier behavior in natural coral-reef habitat. As a result, mortality was 5 to 9 times higher from predation than controls, with mortality increasing with increasing CO_2 (Munday et al. 2010).

CO_2-enriched seawater was far more toxic to eggs and larvae of the seabream, *Pagrus major*, than HCl-acidified seawater when tested at the same seawater pH (Ishimatsu et al. 2004). Accordingly, data on the effects of acidified seawater are suspect when used to estimate the toxicity of CO_2 to marine fishes. Adult yellowtail, *Seriola quinqueradiata*, and olive flounder, *Paralichthys olivaceous*, died within 8 h and 48 h, respectively during exposure to seawater containing 50,000 mg/L CO_2. But only 20% of the starspotted dogfish, *Mustelus manazo*, an elasmobranch, died at 70,000 mg/L CO_2 within 72 h. During exposure to 50,000 mg/L CO_2, olive flounders died after arterial pH decreased but had returned to normal. Exposure to 50,000 mg/L CO_2 rapidly depressed the cardiac output of amberjacks although 10,000 mg/L had no effect; both levels of CO_2 had no effect on blood oxygen levels. Cardiac failure is believed to be responsible for the lethal action of CO_2 on fish (Ishimatsu et al. 2004). Fish have been shown to maintain their oxygen consumption under elevated CO_2 conditions in spite of additional energetic costs incurred by higher CO_2 (Ishimatsu et al. 2008). Authors recommend additional research on possible acclimatization of marine fish to high CO_2 environments, endocrine responses to prolonged CO_2 exposure, and indirect influences through food availability and quality on fish growth, survival and reproduction (Ishimatsu et al. 2008).

Cardiorespiratory responses of cannulated white sturgeon, *Acipenser transmontanus*, were monitored during normocapnic and hypercapnic water conditions (Crocker et al. 2000). Hypercapnia produced: increases in arterial CO_2, ventilatory frequency, and plasma concentrations of cortisol and epinephrine; decreases in arterial pH and plasma glucose; and no change in hematocrit, lactate, or norepinephrine. Hypercapnia increased cardiac output by 22%, arterial pressure by 8%, and heart rate by 8%, but gut blood flow was unaffected. Results indicate that environmental hypercapnia—as is currently experienced at various white sturgeon hatcheries—elicits stress responses that significantly elevate cardiovascular and ventilatory activity levels (Crocker et al. 2000). Sperm motility of white sturgeon was inhibited at pH values of less than 7.5 with maximum motility near pH 8.2 (Ingermann et al. 2002).

Juveniles from three species of commercially-important migratory fish avoided acidified seawater caused by acid sulfate soil discharges near pH 5.0 (Kroon 2005). Avoidance behavior of yellowfin bream *Acanthopagrus australis*, snapper *Pagrus aurata*, and Australian bass *Macquaria novemaculeata* all increased with decreasing pH, with avoidance beginning at pH 7.5 for bream and snapper and pH 6.9 for Australian bass. Author concludes that chronic acid discharges create barriers to migration, potentially affecting the migration of fish to nursery and spawning grounds, with adverse effects on recruitment and stock size (Kroon 2005).

Studies with juveniles of the Japanese whiting, *Sillago japonica* showed some tolerance to slowly-increasing CO_2 concentrations vs. sudden increases (Kikkawa et al. 2006). In that study, step-wise increases from 380 mg CO_2/L to 7,000 mg CO_2/L over a period of 20 hours was fatal to 15% after 18 h. In contrast, a one-step increase from 380 to 7,000 mg CO_2/L was fatal to all surviving fish within a few minutes, suggesting reexamination of the proposed deep-sea disposal of liquified carbon dioxide (Kikkawa et al. 2006). Transportation of juvenile yellowtail kingfish, *Seriola lalandi*, from hatchery to on-growing operations in New Zealand exposes the fish to significantly elevated CO_2 concentrations for about 5 hours (Moran et al. 2008). Blood chemistry, metabolism, and survival were relatively unaffected by 31 hours post-exposure, suggesting that this species can cope with acute exposure to hypercapnia during live transport (Moran et al. 2008).

Juvenile spotted wolffish, *Anarhichas minor*, were subjected for 10 weeks to seawater acidified (from pH 8.1 = controls to 6.98; 6.71; and 6.45) by various concentrations of CO_2 and examined for effects on growth, food conversion and signs of calcareous deposits in the kidney (Foss et al. 2003). Growth, daily feeding rates, and total food consumption were significantly reduced only at the lowest pH tested, whereas food conversion efficiency did not vary significantly between groups. Plasma chloride levels decreased with increasing CO_2. Nephrocalcinosis was observed in all groups, but was most pronounced in the medium and high CO_2 groups, *viz.*, at pH 6.71 and 6.45 (Foss et al. 2003). Juveniles of Atlantic salmon, *Salmo salar*, were exposed to four different levels of CO_2 for 43 days in seawater at pH levels of 7.88 (controls), 7.00, 6.62, and 6.37, respectively (Fivelstad et al. 1998). At pH 7.00, no significant differences from controls were measured in blood chemistry (hematocrit, plasma chloride, plasma sodium) or growth (weight, length, condition factor), and authors concluded that this may be the lowest safe level for salmon smolts pending further temperature-oxygen modifications. At pH 6.62, plasma chloride was significantly reduced. At pH 6.37, plasma sodium and pH were significantly increased, and plasma chloride, oxygen consumption, weight , length, and condition factor were all significantly reduced. Mortality was negligible in all groups and nephrocalcinosis was not observed in any group (Fivelstad et al. 1998). Otoliths (aragonite ear bones) of juvenile fish grown under high CO_2 (low pH) conditions are larger than normal, suggesting that CO_2 moves freely around the otoliths in young fish accelerating otolith growth while the local pH is controlled (Checkley et al. 2009).

Juveniles (15–80 g) of Atlantic cod, *Gadus morhua*, were subjected for 55 days to hypercapnic regimes of 100, 380, or 850 mg/L CO_2 (Moran and Stottrup 2010). Weight gain, growth rate, and condition factor were substantially reduced with increasing CO_2 dosage. The size specific growth trajectories of fish reared under the medium and high CO_2 treatments, when compared to the low treatment, were about 2.5 and 7.5 times lower, respectively. Size variance and mortality rate was the same among treatments, suggesting that the CO_2 levels tested were within the adaptive capacity of this species (Moran and Stottrup 2010). Hypercapnia (pH lowering due to CO_2) protects Atlantic cod against copper intoxication (Larsen et al. 1997). In that study, cod weighing 230–525 g were subjected to one of the following treatments: 9,900 mg/L of CO_2 for 24 h wherein water pH declined from 8.0 (control) to 7.0; 0.4 mg/L Cu for 48 h; and hypercapnia plus copper for 72 h. Cod subjected to hypercapnia alone showed extracellular respiratory acidosis which returned to normal within 12 to 24 h via a chloride-mediated increase in extracellular HCO_3^-. Exposure to copper at pH 8.0 caused a large and progressive increase in plasma Na^+ and Cl^-. Exposure to Cu and CO_2 was associated with smaller elevations of plasma sodium and chloride, showing that hypercapnia had a protective effect on copper-induced osmoregulatory disturbances (Larsen et al. 1997). Most teleost fish are able to fully compensate acid-base disturbances in short-term experiments lasting a few hours to several days (Melzner et al. 2009). Long-term incubation of 4 and 12 months of Atlantic cod to 300 or 600 mg/L CO_2 showed that locomotor performance is not compromised by the different levels of chronic hypercapnia. However, the group exposed for 12 months to 600 mg/L CO_2 had a 2-fold elevation in Na^+/K^+-ATPase protein expression and a 2-fold elevation in Na^+/K^+-ATPase activity; no such elevation was observed in the 300 mg/L group. These results suggest an adjustment of enzymatic activity to cope with the CO_2 induced acid-base load at 600 mg/L CO_2, while under milder hypercapnic conditions Na^+/K^+-ATPase capacity might still be sufficient to maintain acid-base status (Melzner et al. 2009).

Seawater was acidified from 8.18 to 7.01, 6.41 or 6.18 by the addition of 1%, 3%, or 5% CO_2 (Hayashi et al. 2004). Adults of olive flounder, *Paralichthys olivaceus*, yellowtail, *Seriola quinqueradiata*, and starspotted dogfish. *Mustelus manazo*, were exposed for varying intervals. At pH 6.18, all yellowtail died within 8 hours, all flounder within 48 hours, but only 20% of the dogfish died in 72 hours Arterial pH at pH 7.01 and 6.41 initially decreased but was normal within 24 hours. Acid-base regulatory mechanisms differ between teleosts and elasmobranchs, as well as the intensity of acidic stress (Hayashi et al. 2004). Cardiorespiratory responses of yellowtail in seawater equilibrated with 1% or 5% CO_2 were monitored over a period of 72 hours at 20°C (Lee et al. 2003). All fish exposed to the 5% treatment died within 8 hour while none died at the 1% treatment, as expected. No cardiovascular variable measured (cardiac output, heart rate, stroke volume arterial blood pressure) was significantly changed from pre-exposure values during exposure to 1% CO_2. However, arterial CO_2 partial pressure increased, reaching a new steady-state level after 3 hours; arterial blood pH decreased initially but was subsequently restored by elevation of plasma bicarbonate. In contrast, exposure to 5% CO_2 reduced cardiac output through decreasing stroke volume; blood pressure was transiently elevated followed by a precipitous fall before death; blood pH was incompletely restored despite a significant increase in bicarbonate ion; and arterial O_2 partial pressure decreased only shortly before death whereas oxygen content (CaO_2) remained elevated due to a large increase in hematocrit. Cardiac failure is the main cause of death of fish subjected to high environmental CO_2 pressures (Lee et al. 2003).

Gilthead seabream, *Sparus aurata*, held in seawater of pH 7.3 (vs. 8.05 for controls) for 10 days were analyzed for changes in acid-base status and enzyme profiles of muscle and heart (Michaelidis et al. 2007). Experimentals showed a reduction in blood plasma pH and in intracellular pH. Compensation of the acidosis occurred through increased plasma and cellular bicarbonate levels. Lactic dehydrogenase increased and citrate synthase decreased in muscle, and this reflected a shift from aerobic to anaerobic pathways of substrate oxidation. Consequences for slow processes such as growth and reproduction potential are uncertain (Michaelidis et al. 2007).

In tidal salt marshes of South Carolina, spot, *Leiostomus xanthurus*, and mummichogs, *Fundulus heteroclitus*, frequently encounter low oxygen and high carbon dioxide levels. Depending on the stage of tide and the time of day (Cochran and Burnett 1996). Respiratory responses were unaffected over a wide range of oxygen and carbon dioxide concentrations, although whole body lactate concentrations were lower in spot (Cochran and Burnett 1996). Elevated CO_2 may limit the integrity of Antarctic fishes owing to a depression in protein anabolism (Langenbuch and Portner 2003). In that study, isolated hepatocytes from two Antarctic fishes, *Pachycara brachycephalum* and *Lepidonotothen kempi* from cold, deep waters were subjected for 50 minutes to CO_2-induced pH changes in extracellular pH from 7.90 to 6.50. Low pH was associated with a reduction in aerobic metabolic rate of 34–37% and inhibition of protein biosynthesis by about 80% under conditions of severe acidosis in hepatocytes from both species. A parallel drop in intracellular pH probably mediates this effect (Langenbuch and Portner 2003).

Seawater pH affects the toxicity of various organochlorine and organophosphorus insecticides to the mummichog, *Fundulus heteroclitus*, an estuarine cyprinodontiform teleost (Eisler 1970). The toxicity of malathion and phosdrin, two organophosphorus insecticides, were both inversely proportional to pH in the range of 5.5 to 10.0, being most toxic at 5.5–6.2, least toxic in the pH range 9.1–10.0 and intermediate at pH 7.1–8.5. The organochlorine insecticides tested (DDT, methoxychlor, heptachlor) were least toxic at estuarine and oceanic pH values (7.1–8.5) and most toxic under acidic (5.5–6.2) and alkaline pH (9.1–10.1) regimens (Eisler 1970). Frequency of cadmium-induced kidney and intestinal lesions in mummichogs were least under conditions of low pH (Gardner and Yevich 1969). In acute toxicity bioassays with aluminum and striped bass, *Morone saxatilis*, a concentration of 400.0 mg Al^{3+}/L killed all larvae and juveniles in 7 days at pH 7.2; however, all survived 150.0–200.0 mg/L during this same time period (Hall et al. 1985; Buckler et al. 1987). Many additional references on pH effects on fish are listed by Daye (1980).

5.4.2 Reptiles

The ventilatory response of adult and juvenile green turtles, *Chelonia mydas*, to inspired CO_2 (air containing 4,500- to 6,000 ppm CO_2) showed that all animals tested responded with increased ventilation, primarily due to increased respiratory frequency (Jackson et al.

1979). Increased ventilation was noted at increased temperatures in the range of 15° to 35°C (Jackson et al. 1979), suggesting that foraging dives in warm water at elevated air CO_2 levels are of shorter duration and that food intake will be reduced.

Ventilation and gas exchange patterns were examined during voluntary dives in the loggerhead sea turtle. *Caretta caretta,* and contrasted with the changes that occurred during forced submergence (Lutcavage and Lutz 1991). Loggerheads typically spend 86% of their time submerged, with mean dive length of 16.1 minutes. During voluntary dives, oxygen levels decreased, CO_2 increased, and pH held nearly constant. The changes in acid-base status during voluntary submergence differed substantially from those forcibly submerged and may account for the apparently reduced submersion endurance of sea turtles accidentally captured in shrimp trawls (Lutcavage and Lutz 1991). No studies were found on the influence of pH on metals toxicokinetics in marine reptiles (Linder and Grillitsch 2000).

5.4.3 Birds

Fish-eating birds risk increased exposure to dietary methylmercury in acidified habitats, and mercury concentrations in prey may reach levels known to cause reproductive impairment (Scheuhammer 1991). Piscivores do not risk increased exposure to dietary cadmium, lead, or aluminum because these metals are either not increased in fish due to acidification, or the increases are of minor toxicological significance. However, these species may experience a decrease in the availability of dietary calcium due to the pH-related extinction of high-calcium aquatic invertebrate taxa such as molluscs and crustaceans. Decreased availability of dietary calcium is known to adversely affect egg laying and eggshell integrity. Herbivores may risk increased exposure to aluminum and lead, and perhaps cadmium, in acidified environments because certain macrophytes accumulate high concentrations of these metals under acidic conditions (Scheuhammer 1991).

5.4.4 Mammals

Marine mammals demonstrate extreme tolerances to arrested breathing as evidenced by their ability to endure extended breath-hold dives (Kohin et al. 1999). For example, northern elephant seals (*Mirounga angustirostris*) and southern elephant seals (*Mirounga leonina*) routinely dive for 30 minutes, with record dives lasting up to 120 minutes. Northern elephant seals (age 7 to 300 days) subjected to 7% CO_2 showed increases in metabolism up to 38%, increases in heart rate, and suspended respiration—all independent of age. Elevated metabolism during hypercapnia exposure is attributable to suspended respiration. In young elephant seals, metabolic down-regulation is not an automatic protective response to experimentally-imposed hypercapnia (Kohin et al. 1999).

Fish-eating mammals may be at risk from increased exposure to dietary methylmercury in acidified habitats (Scheuhammer 1991). Under acidic conditions, mercury levels in prey may reach levels known to cause reproductive impairment, but this not the case for aluminum, lead, or cadmium. Piscivores may experience decreased growth of neonates owing to a decrease in dietary calcium from pH-related extinction of various high-calcium invertebrate taxa. Herbivores may experience increased risk from dietary exposure of macrophytes that accumulate aluminum, lead, and cadmium (Scheuhammer 1991).

5.5 Literature Cited

Abbott, O.J. 1977. The toxicity of ammonium molybdate to marine invertebrates, *Mar. Pollut. Bull.*, 8, 204–205.

Adey, W.H. 1998. Coral reefs: algal structured and mediated ecosystems in shallow, turbulent, alkaline waters, *J. Phycol.*, 34, 393–406.

Albright, R., B. Mason, M. Miler, and C. Langdon. 2010. Ocean acidification compromises recruitment success of the threatened Caribbean coral *Acropora palmata*, *Proc. Natl. Acad. Sci. USA*, 107, 20400–20404.

Allan, G.L. and G.B. Maguire. 1992. Effects of pH and salinity on survival, growth and osmoregulation in *Penaeus monodon* Fabricius, *Aquaculture*, 107, 33–47.

Anthony, K.R.N, D.J. Kline, G. Diaz-Pulido, S. Dove, and O. Hoegh-Guldberg. 2008. Ocean acidification causes bleaching and productivity loss in coral reef builders, *Proc. Natl. Acad. Sci. USA*, 105, 17442–17446.

Arnold, K.E., H.S. Findlay, J.I. Spicer, C.L. Daniels, and D. Boothroyd. 2009. Effect of CO_2-related acidification on aspects of the larval development of the European lobster, *Homarus gammarus* (L.), *Biogeosciences*, 6, 1747–1754.

Balch, W.M. and V.J. Fabry. 2008. Ocean acidification: documenting its impact on calcifying phytoplankton at basin scales, *Mar. Ecol. Prog. Ser.*, 373, 239–247.

Bamber, R.N. 1987. The effects of acidic sea water on young carpet-shell clams *Venerupis decussata* (L.) (Mollusca: Veneracea), *J. Exp. Mar. Biol. Ecol.*, 108, 241–260.

Bamber, R.N. 1990. The effects of acidic seawater on three species of lamellibranch molluscs, *J. Exp. Mar. Biol. Ecol.*, 143, 181–191.

Barker, S. and H. Elderfield. 2002. Foraminiferal calcification response to glacial-interglacial changes in atmospheric CO_2, *Science*, 297, 833–836.

Batten, S.D. and R.N. Bamber. 1996. The effects of acidified seawater on the polychaete *Nereis virens* Sars, 1835, *Mar. Pollut. Bull.*, 32, 283–287.

Beer, S. and E. Koch. 1996. Photosynthesis of marine macroalgae and seagrasses in globally changing CO_2 environments, *Mar. Ecol. Prog. Ser.*, 141, 199–204.

Beesley, A., D.M. Lowe, C.K. Pascoe, and S. Widdicombe. 2008. Effects of CO_2-induced seawater acidification on the health of *Mytilus edulis*, *Climate Res.*, 37, 215–225.

Bellerby, R.G.J., K.G. Schulz, U. Riebesell, C. Neill, G. Nondal, E. Heegaard, T. Johannessen, and K.R. Brown. 2008. Marine ecosystem community carbon and nutrient uptake stoichiometry under varying ocean acidification during the PeECE III experiment, *Biogeosciences*, 5, 1517–1527.

Berge, J.A., B. Bjerking, O. Pettersen, M.T. Schaanning, and S. Oxnevad. 2006. Effects of increased sea water concentrations of CO_2 on growth of the bivalve *Mytilus edulis* L., *Chemosphere*, 62, 681–687.

Bibby, R., P. Cleall-Harding, S. Rundle, S. Widdicombe, and J. Spicer. 2007. Ocean acidification disrupts induced defenses in the intertidal gastropod *Littorina littorea*, *Biol. Lett.*, 3, 699–701.

Bibby, R., S. Widdicombe, H. Parry, J. Spicer, and R. Pipe. 2008. Effects of ocean acidification on the immune response of the blue mussel *Mytilus edulis*, *Aquat. Biol.*, 2, 67–74.

Bijma, J., B. Honisch, and R.E. Zeebe. 2002. Impact of the ocean carbonate chemistry on living foraminiferal shell weight: comment on "Carbonate ion concentration in glacial-age deep waters of the Caribbean Sea" by W.S. Broecker and E. Clark, *Geochem. Geophys. Geosyst.*, 3, (11), 1064, doi:10.1029/2002GC000388, 2002.

Bjork, M., A. Weil, S. Semesi, and S. Beer. 1997. Photosynthetic utilisation of inorganic carbon by seagrasses from Zanzibar, East Africa, *Mar. Biol.*, 129, 363–366.

Blust, R., A. Fontaine, and W. Decleir. 1991. Effect of hydrogen ion and inorganic complexing on the uptake of copper by the brine shrimp *Artemia franciscana*, *Mar. Ecol. Prog. Ser.*, 76, 273–282.

Borowitzka, M.A. 1981. Photosynthesis and calcification in the articulated coralline red algae *Amphiroa anceps* and *A. foliacea*, *Mar. Biol.*, 62, 17–23.

Buckler, D.R., P.M. Mehrle, L. Cleveland, and F.J. Dwyer. 1987. Influence of pH on the toxicity of aluminum and other inorganic contaminants to east coast striped bass, *Water Air Soil Pollut.*, 35, 97–106.

Buddemeier, R.W. and J.R. Ware. 2003. Coral reef decline in the Caribbean, *Science*, 302, 391–392.

Bull, H.O. 1940. Studies on conditioned responses in fishes. IX. Discrimination of changes in hydrogen-ion concentration by marine teleosts, *Rep. Dove Mar. Lab, Ser., 3, 7*, 21–31.

Burgents, J.E., K.G. Burnett, and L.E. Burnett. 2005. Effects of hypoxia and hypercapnic hypoxia on the localization and the elimination of *Vibrio campbelli* in *Litopenaeus vannamei*, the Pacific white shrimp, *Biol. Bull., 208*, 159–168.

Burkhardt, S., I. Zondervan, and U. Riebesell. 1999. Effect of CO_2 concentration on C:N:P ratio in marine phytoplankton: a species comparison, *Limnol. Ocean,* 44, 683–690.

Burnett, L., N. Terwilliger, A. Carroll, D. Jorgensen, and D. Scholnick. 2002. Respiratory and acid-base physiology of the purple sea urchin, *Strongylocentrotus purpuratus*, during air exposure: presence and function of a facultative lung, *Biol. Bull., 203*, 42–50.

Calabrese, A. and H.C. Davis. 1966. The pH tolerance of embryos and larvae of *Mercenaria mercenaria* and *Crassostrea virginica, Biol. Bull., 131*, 427–436.

Carter, L. 1964. Effects of acidic and alkaline effluents on fish in seawater, *Effluent Water Treat. J., 4*, 484–486.

Cecchini, S., M. Saroglia, G. Caricato, G. Terova, and L. Sileo. 2001. Effects of graded environmental hypercapnia on sea bass (*Dicentrarchus labrax* L.) feed intake and acid-base balance, *Aquacult. Res., 32*, 499–502.

Checkley, D.M. Jr., A.G. Dickson, M. Takahashi, J.A. Radich, N. Eisenkolb, and R. Asch. 2009. Elevated CO_2 enhances otolith growth in young fish, *Science*, 324, 1683.

Clapper, D.L., J.A. Davis, P.J. Lamothe, C. Patton, and D. Epel. 1985a. Involvement of zinc in the regulation of pH, motility, and acrosome reactions in sea urchin sperm, *J. Cell. Biol., 100*, 1817–1824.

Clapper, D.L., P.J. Lamothe, J.A. Davis, and D. Epel. 1985b. Sperm motility in the horseshoe crab. V: zinc removal mediates chelator initiation of motility, *J. Exp. Zool., 236*, 83–91.

Clark, D., M. Lamare, and M. Barker. 2009. Response of sea urchin pluteus larvae (Echinodermata:Echinoidea) to reduced seawater pH: a comparison among a tropical, temperate, and a polar species, *Mar. Biol., 156*, 1125–1137.

Cochran, R.E. and L.E. Burnett. 1996. Respiratory responses of the salt marsh animals, *Fundulus heteroclitus, Leiostomus xanthurus,* and *Palaemonetes pugio* to environmental hypoxia and hypercapnia and to the organophosphate pesticide, azinphosmethyl, *J. Exp. Mar. Biol. Ecol., 195*, 125–144.

Cohen, A.L., D.C. McCorkle, S. De Putron, G.A. Gaetani, and K.A. Rose. 2009. Morphological and compositional changes in the skeletons of new coral recruits reared in acidified seawater: insights into the biomineralization response to ocean acidification, *Geochem. Geophys. Geosyst., 10*, Q07005, doi:1029/2009GC002411.

Collins, S. and G. Bell. 2004. Phenotypic consequences of 1,000 generations of selection at elevated CO_2 in a green alga, *Nature, 431*, 566–569.

Comeau, S., S. Gorsky, R. Jeffree, J.L. Teyssie, and J.P. Gattuso. 2009. Impact of ocean acidification on a key Arctic pelagic mollusc (*Limacina helicina*), *Biogeosciences, 6*, 1877–1882.

Connell, D.B., J.G. Sanders, G.F. Riedel, and G.R. Abbe. 1991. Pathways of silver uptake and trophic transfer in estuarine organisms, *Environ. Sci. Technol.*, 25, 921–924.

Cooley, S.R. and S.C. Doney. 2009. Anticipating ocean acidification's economic consequences for commercial fisheries, *Environ. Res. Lett.*, 024007; doi:10.1088/1748–9326/4/02407.

Crocker, C.E., A.P. Farrell, A.K. Gamperl, and J.E. Cech, Jr. 2000. Cardiorespiratory responses of white sturgeon to environmental hypercapnia, *Amer. J. Physiol. Regul. Integr. Comp. Physiol.*, 279, R617–R628.

Czerny, J., J.B. e Ramos, and U. Riebesell. 2009. Influence of elevated CO_2 concentrations on cell division and nitrogen fixation rates in the bloom-forming cyanobacterium *Nodularia spumigena*, *Biogeosciences*, 6, 1865–1875.

Dashfield, S., P.J. Somerfield, S. Widdicombe, M.C. Austen, and M. Nimmo. 2008. Impacts of ocean acidification and burrowing urchins on within-sediment pH profiles and subtidal nematode communities, *J. Exp. Mar. Biol. Ecol.*, 365, 46–52.

Davies, J.K. 1991. Reactions of sand smelt to low pH sea-water, *Mar. Pollut. Bull.*, 22, 74–77.

Daye, P.G. 1980. Effects of ambient pH on fish: an annotated bibliography, *Canad. Tech. Rep. Fish. Aquat. Sci.*, 950, 1–28.

Delille, B., J. Harlay, I. Zondervan, S. Jacquet, L. Chou, R. Wollast, R.G.J. Bellerby, M. Frankignoulle, A.V. Borges, U. Riebesell, and J.P. Gattuso. 2005. Response of primary production and calcification to changes of pCO_2 during experimental blooms of the coccolithophorid *Emiliana huxleyi*, *Global Biogeochem. Cycles*, 19, GB2023, doi:10.1029/2004GB002318.

Dong, Q., B. Eudeline, S.K. Allen, and T.R. Tiersch Jr. 2002. Factors affecting sperm motility of tetraploid Pacific oysters, *J. Shellfish Res.*, 21, 719–723.

Duke, R.Q. 1967. Possible routes of zinc-65 from an experimental estuarine environment to man, *J. Water Pollut. Contr. Feder.*, 39, 536–542.

Dupont, S., J. Havenhand, W. Thorndyke, L. Peck, and M. Thorndyke. 2008. Near-future level of CO_2-driven ocean acidification radically affects larval survival and development in the brittlestar *Ophiothrix fragilis*, *Mar. Ecol. Prog. Ser.*, 373, 285–294.

Dupont, S., B. Lundve, and M. Thorndyke. 2010a. Near future ocean acidification increases growth rate of the lecithotrophic larvae and juveniles of the sea star *Crossaster papposus*, *J. Exp. Zool.*, 314B, 1–8.

Dupont, S., O. Ortega-Martinez, and M. Thorndyke. 2010b. Impact of near-future ocean acidification on echinoderms, *Ecotoxicology*, 19, 449–462.

Edmunds, P.J. 2007. Evidence for a decadal-scale decline in the growth of juvenile scleractinian corals, *Mar. Ecol. Prog. Ser.*, 341, 1–13.

Egilsdottir, H., J.I. Spicer, and S.D. Rundle. 2009. The effect of CO_2 acidified sea water and reduced salinity on aspects of the embryonic development of the amphipod *Echinogammarus marinus* (Leach), *Mar. Pollut. Bull.*, 58, 1187–1191.

Eisler, R. 1970. Factors affecting pesticide-induced toxicity in an estuarine fish, *U.S. Bureau Sport Fish Wild.* Tech. Paper, 45, 1–20.

Eisler, R. 1981. *Trace Metal Concentrations in Marine Organisms*, Pergamon Press, Elmsford, New York, 79–87.

Ellis, R.P., J. Bersey, S.D. Rundle, J.M. Hall-Spencer, and J.I. Spicer. 2009. Subtle but significant effects of CO_2 acidified seawater on embryos of the intertidal snail, *Littorina obtusata*, *Aquat. Biol.*, 5, 41–48.

Engel, A., I. Zondervan, K. Aerts, L. Beaufort, A. Benthien, L. Chou, B. Delille, J.P. Gattuso, J. Harlay, C. Hermann, L. Hoffman, S. Jacquet, J. Nejstgaard, M.D. Pizay, E. Rochelle-Newall, U. Schneider, A. Terbrueggen, and U. Riebesell. 2005. Testing the direct effect of CO_2 concentration on a bloom of the coccolithophorid *Emiliana huxleyi* in mesocosm experiments, *Limnol. Ocean*, 50, 493–507.

Fabry, V.J. 2008. Marine calcifiers in an a high-CO_2 ocean, *Science*, 320, 1020–1022.

Fabry, V.J., B.A. Seibel, R.A. Feely, and J.C. Orr. 2008. Impacts of ocean acidification on marine fauna and ecosystem processes, *ICES J. Mar. Sci.*, 65, 414–432.

Feely, R.A., S.C. Doney, and S.R. Cooley. 2009. Ocean acidification. Present conditions and future changes in a high-CO_2 world, *Oceanography*, 22, 36–47.

Findlay, H.S., M.A. Kendall, J.I. Spicer, and S. Widdicombe. 2009. Future high CO_2 in the intertidal may compromise adult barnacle *Semibalanus balanoides* survival and embryonic development rate, *Mar. Ecol. Prog. Ser.*, 389, 193–202.

Findlay, H.S., M.A. Kendall, J.I. Spicer, and S. Widdicombe. 2010. Relative influences of ocean acidification and temperature on intertidal barnacle post-larvae at the northern edge of their geographic distribution, *Estuar. Coast. Shelf Sci.*, 86, 675–682.

Findlay, H.S., M.A. Kendall, J.I. Spicer, and S. Widdicombe. 2010a. Post-larval development of two intertidal barnacles at elevated CO_2 and temperature, *Mar. Biol.*, 157, 725–735.

Fine, M. and D. Tchernov. 2007. Scleractinian coral species survive and recover from decalcification, *Science*, 315, 1811.

Fivelstad, S., H. Haavik, F. Lovik, and A.B. Olsen. 1998. Sublethal effects and safe levels of carbon dioxide in seawater for Atlantic salmon postsmolts (*Salmo salar* L.): ion regulation and growth, *Aquaculture*, 160, 305–316.

Foss, A., B.A. Rosnes, and V. Oiestad. 2003. Graded environmental hypercapnia in juvenile spotted wolffish (*Anarhichas minor* Olafsen): effects on growth, food conversion efficiency and nephrocalcinosis, *Aquaculture*, 220, 607–617.

Fu, F.X., M.R. Mulholland, N.S. Garcia, A. Beck, P.W. Bernhardt, M.E. Warner, S.A. Sanudo-Wilhelmy, and D.A. Hutchins. 2008. Interactions between changing pCO_2, N_2 fixation, and Fe limitation in the marine unicellular cyanobacterium *Crocosphaera*, *Limnol. Ocean.*, 53, 2472–2484.

Fu, F.X., M.E. Warner, Y. Zhang, Y. Feng, and D.A. Hutchins. 2007. Effects of increased temperature and CO_2 on photosynthesis, growth, and elemental ratios in marine *Synechococcus* and *Prochlorococcus* (cyanobacteria), *J. Phycol.*, 43, 485–496.

Gambrell. R.P., V.R. Collard, C.N. Reddy, and W. H. Patrick Jr. 1977. Trace and toxic metal uptake by marsh plants as affected by Eh, pH., and salinity, *U.S. Army Eng. Water. Exp. Sta. Vicksburg, Mississippi*, Dredged Mat. Res. Prog. Tech. Rept., D-77–40, 1–124.

Gao, K., Y. Aruga, K. Asada, T. Ishihara, T. Akano, and M. Kiyohara. 1993a. Calcification in the articulated coralline alga *Corallina pilulifera*, with special reference to the effect of elevated CO_2 concentration, *Mar. Biol.*, 117, 129–132.

Gao, K., Y. Aruga, K. Asada, and M. Kiyohara. 1993. Influence of enhanced CO_2 on growth and photosynthesis of the red algae *Gracilaria* sp. and *G. chilensis, J. Appl. Phycol.*, 5, 563–571.

Gardner, G.R. and P.P. Yevich. 1969. Toxicological effects of cadmium on *Fundulus heteroclitus* under various pH, oxygen, salinity and temperature regimes, *Amer. Zool.*, 9, 1096.

Gardner, T.A., I.M. Cole, J.A. Gill, A. Grant, and A.R.Watkinson. 2003. Long-term region-wide declines in Caribbean corals, *Science*, 302, 958–960.

Gazeau, F., C. Quiblier, J.M. Jansen, J.P. Gattuso, J.J. Middelburg, and C.R. Heip. 2007. Impact of elevated CO_2 on shellfish calcification, *Geophys. Res. Lett.*, 34, L07603, doi:10.1029/2006GL028554,

Gooding, R.A., C.D.G. Harley, and E. Tang. 2009. Elevated water temperature and carbon dioxide concentrations increase the growth of a keystone echinoderm, *Proc. Natl. Acad. Sci. USA*, 106, 9316–9321.

Gregg, K.W. and C.B. Metz. 1976. Physiological parameters of the sea urchin acrosome reaction, *Biol. Reprod.*, 14, 405–411.

Grice, G.D., P.H. Wiebe, and E. Hoagland. 1973. Acid-iron waste as a factor affecting the distribution and abundance of zooplankton in the New York Bight, *Estuar. Coastal. Mar. Sci.*, 1, 45–50.

Guinotte, J.M., R.W. Buddemeier, and J. Kleypas. 2003. Future coral reef habitat marginality: temporal and spatial effects of climate change in the Pacific basin, *Coral Reefs*, 22, 551–558.

Gutowska, M.A. and F. Melzner. 2009. Abiotic conditions in cephalopod (*Sepia officinalis*) eggs: embryonic development at low pH and high pCO_2, *Mar. Biol.*, 156, 515–519.

Gutowska, M.A., H.O. Portner, and F. Melzner. 2008. Growth and calcification in the cephalopod *Sepia officinalis* under elevated seawater pCO_2, *Mar. Ecol. Prog. Ser.*, 373, 303–309.

Hall, L.W. Jr., A.E. Pinkney, L.O. Horsemen, and S.E. Finger. 1985. Mortality of striped bass larvae in relation to contaminants and water quality in a Chesapeake Bay tributary, *Trans. Amer. Fish. Soc.*, 114, 861–868.

Hampson, M.A. 1967. Uptake of radioactivity by aquatic plants and location in the cells. I. The effect of pH on the Sr-90 and Y-90 uptake by the green alga *Ulva lactuca* and the effect of stable yttrium on Y-90 uptake, *J. Exp. Botany*, 18, 17–53.

Harris, J.O., G.B. Maguire, S.J. Edwards, and S.M. Hindrum. 1999. Effect of pH on growth rate, oxygen consumption rate, and histopathology of gill and kidney tissue for juvenile greenlip abalone, *Haliotis laevigata* Donovan and blacklip abalone, *Haliotis rubra* Leach, *J. Shellfish Res.*, 18, 611–619.

Harris, R.P. 1994. Zooplankton grazing on the coccolithophore *Emiliania huxleyi* and its role in inorganic carbon flux, *Mar. Biol.*, 119, 431–439.

Hauton, G., T. Tyrrell, and J. Williams. 2009. The subtle effects of sea water acidification on the amphipod *Gammarus locusta*, *Biogeosci.*, 6, 1479–1489.

Havenhand, J., F.R. Buttler, M.C. Thorndyke, and J.E. Williamson. 2008. Near-future levels of ocean acidification reduce fertilization success in a sea urchin, *Curr. Biol.*, 18, 651–652.

Havenhand, J.N. and P. Schlegel. 2009. Near-future levels of ocean acidification do not affect sperm motility and fertilization kinetics in the oyster *Crassostrea gigas*, *Biogeosciences*, 6, 3009–3015.

Hayashi, M., J. Kita, and A. Ishimatsu. 2004. Acid-base responses to lethal aquatic hypercapnia in three marine fishes. *Mar. Biol.*, 144, 153–160.

Hein, M. and K. Sand-Jensen.1997. CO_2 increases oceanic primary production, *Nature*, 388, 526–527.

Hendriks, I.E., C.M. Duarte, and M. Alvarez. 2010. Vulnerability of marine biodiversity to ocean acidification: a meta-analysis, *Estuar Coastal Shelf Sci.*, 86, 157–164.

Herfort, L., B. Thake, and I. Taubner. 2008. Bicarbonate stimulation of calcification and photosynthesis in two hermatypic corals, *J. Phycol.*, 44, 91–98.

Hinga, K.R. 2002. Effects of pH on coastal marine phytoplankton, *Mar. Ecol. Prog. Ser.*, 238, 281–300.

Hoegh-Guldberg, O. 1999. Climate change, coral bleaching and the future of the world's coral reefs, *Mar. Freshwat. Res.*, 50, 839–866.

Hoegh-Guldberg, O., P.J. Mumby, A.J. Hooten, R.S. Steneck, P. Greenfield, E. Gomez, C.D. Harvell, P.F. Sale, A.J. Edwards, K. Caldeira, N. Knowlton, C.M. Eakin, R. Iglesias-Prieto, N. Muthiga, R.H. Bradbury, A. Dubi, and M.E. Hatziolos. 2007. Coral reefs under rapid climate change and ocean acidification, *Science*, 318, 1737–1742.

Hofmann, G.E., M.J. O'Donnell, and A.E. Todgham. 2008. Using functional genomics to explore the effects of ocean acidification on calcifying marine organisms, *Mar. Ecol. Prog. Ser.*, 373, 219–225.

Hofmann, M. and H.J. Schellnhuber. 2009. Oceanic acidification affects marine carbon pump and triggers extended marine oxygen holes. *Proc. Natl. Acad. Sci. USA*, 106, 3017–3022.

Huesemann, M.H., A.D. Skillman, and E.A. Crecelius. 2002. The inhibition of marine nitrification by ocean disposal of carbon dioxide, *Mar. Pollut. Bull.*, 44, 142–148.

Hughes, T.P., A.H. Baird, D.R. Bellwood, M. Card, S.R. Connolly, C. Folke, R. Grosberg, O. Hoegh-Guldberg, J.B.C. Jackson, J. Kleypas, J.M. Lough, P. Marshall, M. Nystrom, S.R. Palumbi, J.M. Pandolfi, B. Rosen, and J. Roughgarden. 2003. Climate change, human impacts, and the resilience of coral reefs, *Science*, 301, 929–933.

Hughes, T.P and J.E. Tanner. 2000. Recruitment failure, life histories, and long-term decline of Caribbean corals, *Ecology*. 81, 2250–2263.

Iglesias-Rodriguez, M.D., P.R. Halloran, R.E.M. Rickaby, I.R. Hall, E. Colmeno-Hidalgo, J.R. Gittins, D.R.H. Green, T. Tyrrell, S.J. Gibbs, P. Von Dassow, E. Rehm, E.V. Armbrust, and K.P. Boessenkool. 2008. Phytoplankton calcification in a high-CO_2 world, *Science*, 320 (5874), 336–340.

Ingermann, R.L., M. Holcomb, M.L. Robinson, and J.G. Cloud. 2002. Carbon dioxide and pH affect sperm motility of white sturgeon (*Acipenser transmontanus*), *J. Exp. Biol.*, 205, 2885–2890.

Invers, O., F. Tomas, M. Perez, and J. Romero. 2002. Potential effect of increased global CO_2 availability on the depth distribution of the seagrass *Posidonia oceanica* (L.) Delile: a tentative assessment using a carbon balance model, *Bull. Mar. Sci.*, 71, 1191–1198.

Ishimatsu, A., M. Hayashi, and T. Kikkawa. 2008. Fishes in high-CO_2 acidified oceans, *Mar. Ecol. Prog. Ser.*, 373, 295–302.

Ishimatsu, A., M. Hayashi, K.S. Lee, T. Kikkawa, and J. Kita. 2005. Physiological effects on fishes in a high-CO_2 world, *J. Geophys. Res.*, 110, C09S09, doi:10.1029/2004JC002564.

Ishimatsu, A., T. Kikkawa, M. Hayashi, K.S. Lee, and J.S. Kita. 2004. Effects of CO_2 on marine fish; larvae and adults, *J. Ocean*, 60, 731–741.

Jackson, D.C., D.R. Kraus, and H.D. Prange. 1979. Ventilatory response to inspired CO_2 in the sea turtle: effects of body size and temperature, *Respir. Physiol.*, 38, 71–81.

Jansen, H. and D.A. Wolf-Gladrow. 2001. Carbonate dissolution in copepod guts: a numerical model, *Mar. Ecol. Prog. Ser.*, 221, 199–207.

Jokiel, P.L , K.S. Rodgers, I.B. Kuffner, A.J. Andersson, E.F.R. Cox, and F.T. Mackenzie. 2008. Ocean acidification and calcifying reef organisms: a mesocosm investigation, *Coral Reefs*. 27, 473–483.

Kawaguchi, S., H. Kurihara, R. King, L. Hale, T. Berli, J.P. Robinson, A. Ishida, M. Wakita, P. Virtue, S. Nicol, and A. Ishimatsu. 2010. Will krill fare well under Southern Ocean acidification? *Biol. Lett.*, doi:10.1098/rsbl.2010.0777.

Kikkawa, T., A. Ishimatsu, and J. Kita. 2003. Acute CO_2 tolerance during the early developmental stages of four marine teleosts, *Environ. Toxicol.*, 18, 375–382.

Kikkawa, T., J. Kita, and A. Ishimatsu. 2004. Comparison of the lethal effect of CO_2 and acidification on red sea bream (*Pagrus major*) during the early developmental stages, *Mar. Pollut. Bull.*, 48, 108–110.

Kikkawa, T., T. Sato, J. Kita, and A. Ishimatsu. 2006. Acute toxicity of temporally varying seawater CO_2 conditions on juveniles of Japanese sillago (*Sillago japonica*), *Mar. Pollut. Bull.*, 52, 621–625.

Kikkawa, T., Y. Watanabe, Y. Katayama, J. Kita, and A. Ishimatsu. 2008. Acute CO_2 tolerance limits of juveniles of three marine invertebrates, *Sepia lycidas*, *Sepioteuthis lessoniana*, and *Marsupenaeus japonicus*, *Plankton Benthos Res.*, 3, 184–187.

Kim, J.M., K. Lee, K. Shin, J.H. Kang, H.W. Lee, M. Kim, P.G. Jang, and M.C. Jang. 2006. The effect of seawater CO_2 concentration on growth of a natural phytoplankton assemblage in a controlled mesocosm experiment, *Limnol. Ocean*, 51, 1629–1636.

Kinne, O. and H. Rosenthal. 1967. Effects of sulfuric water pollutants on fertilization, embryonic development and larvae of the herring, *Clupea harengus*, *Mar. Biol.*, 1, 65–83.

Kinsey, D.W. 1978. Alkalinity changes and coral reef calcification, *Limnol. Ocean.*, 23, 989–991.

Kleypas, J.A., R.W. Buddemeier, D. Archer, J.P. Gattuso, C. Langdon, and B.N. Opdyke. 1999. Geochemical consequences of increased atmospheric carbon dioxide on coral reefs, *Science*, 284, 118–120.

Kleypas, J.A. and K.K. Yates. 2009. Coral reefs and ocean acidification, *Oceanography*, 22 (4), 108–117.

Knutzen, J. 1981. Effects of decreased pH on marine organisms, *Mar. Pollut. Bull.*, 12, 25–29.

Kohin, S., T.M. Williams, and C.L. Ortiz. 1999. Effects of hypoxia and hypercapnia on aerobic metabolic processes in northern elephant seals, *Respir. Physiol.*, 117, 59–72.

Kroon, F.J. 2005. Behavioural avoidance of acidified water by juveniles of four commercial fish and prawn species with migratory life stages, *Mar. Ecol. Prog. Ser.*, 285, 193–204.

Kubler, J.E., A.M. Johnston, and J.A. Raven. 1999. The effects of reduced and elevated CO_2 and O_2 on the seaweed *Lomentaria articulata*, *Plant Cell Environ.*, 22, 1303–1310.

Kuffner, I.B., A.J. Andersson, P.L. Jokiel, K.S. Rodgers, and F.T. Mackenzie. 2008. Decreasing abundance of crustose coralline algae due to ocean acidification, *Nature Geosci.*, 1, 114–117.

Kurihara, H. 2008. Effects of CO_2-driven ocean acidification on the early developmental stages of invertebrates, *Mar. Ecol. Prog. Ser.*, 373, 275–284.

Kurihara, H. and A. Ishimatsu. 2008. Effects of high CO_2 seawater on the copepod (*Acartia tsuensis*) through all life stages and subsequent generations, *Mar. Pollut. Bull.*, 56, 1086–1090.

Kurihara, H., S. Kato, and A. Ishimatsu. 2007. Effects of increased seawater pCO_2 on early development of the oyster *Crassostrea gigas*, *Aquat. Biol.*, 1, 91–98.

Kurihara, H., M. Matsui, H. Furukawa, M. Hayashi, and A. Ishimatsu. 2008. Long-term effects of predicted future seawater CO_2 conditions on the survival and growth of the marine shrimp *Palaemon pacificus*, *J. Exp. Mar. Biol. Ecol.*, 367, 41–46.

Kurihara, H., S. Shimode, and Y. Shirayama. 2004a. Sub-lethal effects of elevated concentrations of CO_2 on planktonic copepods and sea urchins, *J. Ocean*, 60, 743–750.

Kurihara, H., S. Shimode, and Y. Shirayama. 2004. Effects of raised CO_2 concentration on the egg reproduction rate and early development of two marine copepods (*Acartia steueri* and *Acartia erythraea*), *Mar. Pollut Bull.*, 49, 721–727.

Kurihara, H. and Y. Shirayama. 2004. Effects of increased atmospheric CO_2 on sea urchin early development, *Mar. Ecol. Prog. Ser.*, 274, 161–169.

Kuwatani, Y. and T. Nishii. 1969. Effects of pH of culture water on the growth of the Japanese pearl oyster, *Nippon Suisan Gakkaishi (Japan. Soc. Fish. Sci.)*, 35, 342–350.

Langdon, C., W.S. Broecker, D.E. Hammond, E. Glenn, K. Fitzsimmons, S.G. Nelson, T.K. Peng, I. Hajdas, and G. Bonani. 2003. Effect of elevated CO_2 on the community metabolism of an experimental coral reef, *Global Biogeochem. Cycles*, 17(1), 1011, doi:10.1029/2002GB001941,2003.

Langenbuch, M. and H.O. Portner. 2003. Energy budget of hepatocytes from Antarctic fish (*Pachycara brachycephalum* and *Lepidonotothen kempi*) as a function of ambient CO_2: pH-dependent limitations of cellular protein synthesis?, *J. Exp. Bio.*, 206, 3895–3903.

Langenbuch, M. and H.O. Portner. 2004. High sensitivity to chronically elevated CO_2 levels in a eurybenthic marine sipunculid, *Aquat. Toxicol.*, 70, 55–61.

Langer, G., M. Geisen, K.H. Baumann, J. Klas, U. Riebesell, S. Thoms, and J.R. Young. 2006. Species-specific responses of calcifying algae to changing seawater carbonate chemistry, *Geochem. Geophys. Geosyst.* 7. Q09006, doi:10.1029/2005GC001227.

Lannig, G., S. Eilers, H.O. Portner, I.M. Sokolva, and C. Bock. 2010. Impact of ocean acidification on energy metabolism of oyster, *Crassostrea gigas*–changes in metabolic pathways and thermal response, *Mar. Drugs*, 8, 2318–2339.

Larsen, B.K., H.O. Portner, and F.B. Jensen. 1997. Extra- and intracellular acid-base balance and ionic regulation in cod (*Gadus morhua*) during combined and isolated exposures to hypercapnia and copper, *Mar. Biol.*, 128, 337–346.

Leclercq, N., J.P. Gattuso, and J. Jaubert. 2002. Primary production, respiration, and calcification of a coral reef mesocosm under increased CO_2 partial pressure, *Limnol. Ocean*, 47, 558–564.

Lee, K.S., J. Kita, and I. Ishimatsu. 2003. Effects of lethal levels of environmental hypercapnia on cardiovascular and blood-gas status in yellowtail, *Seriola quinqueradiata*, *Zool. Sci.* (Japan), 20, 417–422.

Linder, G. and B. Grillitsch. 2000. Ecotoxicology of Metals. In D.W. Sparling, G. Linder, and C.A. Bishop (Eds.). *Ecotoxicology of Amphibians and Reptiles.* SETAC Press, Pensacola, Florida, pp 324–458.

Lough, J.M. 2008. 10th anniversary review: a changing climate for coral reefs, *J. Environ. Monitor.*, 10, 21–29.

Lu, Z., N. Jiao, and H. Zhang. 2006. Physiological changes in marine picocyanobacterial *Synechococcus* strains exposed to elevated CO_2 partial pressure, *Mar. Biol. Res.*, 2, 424–430.

Lutcavage, M.E. and P.L. Lutz. 1991. Voluntary diving metabolism and ventilation in the loggerhead sea turtle, *J. Exp. Mar. Biol. Ecol.*, 147, 287–296.

Maier, C., J. Hegeman, M.G. Weinbauer, and J.P. Gattuso. 2009. Calcification of the cold-water coral *Lophelia pertusa* under ambient and reduced pH, *Biogeosciences Disc.*, 6, 1875–1901.

Manzello, D.P., J.A. Kleypas, D.A. Budd, C.M. Eakin, P.W. Glynn, and C. Langdon. 2008. Poorly cemented coral reefs of the eastern tropical Pacific: possible insights into reef development in a high CO_2 world, *Proc. Natl. Acad. Sci. USA*, 105, 14050–10455.

Martin, S. and J.P. Gattuso. 2009. Response of Mediterranean coralline algae to ocean acidification and elevated temperature, *Global Change Biol.*, 15, 2089–2100.

Martin, S., R. Rodolfo-Metalpa, E. Ransome, S. Rowley, M.C. Buia, J.P. Gattuso, and J. Hall-Spencer. 2008. Effects of naturally acidified seawater on seagrass calcareous epibionts, *Biol. Lett.*, 4, 689–692.

Marubini, F. And M.J. Atkinson. 1999. Effects of lowered pH and elevated nitrate on coral calcification, *Mar. Ecol. Prog. Ser.*, 188, 117–121.

Marubini, F., C. Ferrier-Pages, and J.P. Cuif. 2002. Suppression of skeletal growth in scleractinian corals by decreasing ambient carbonate-ion concentration: a cross-family comparison, *Proc. Roy. Soc. Lond. B*, 270, 179–184.

Mayor, D.J., C. Matthews, K. Cook, A.F. Zuur, and S. Hay. 2007. CO_2-induced acidification affects hatching success in *Calanus finmarchicus*, *Mar. Ecol. Prog. Ser.*, 350, 91–97.

McDonald, M.R., J.B. McClintock, C.D. Amsler, D. Rittschoff, R.A. Angus, B. Orihuela, and K. Lutostanski. 2009. Effects of ocean acidification over the life history of the barnacle *Amphibalanus amphitrite, Mar. Ecol. Prog. Ser.,* 385, 179–187.

Melzner, F., S. Gobel, M. Langenbuch, M.A. Gutowska, H.O. Portner, and M. Lucassen. 2009. Swimming performance in Atlantic cod (*Gadus morhua*) following long-term (4–12 months) acclimation to elevated seawater pCO_2, *Aquat. Toxicol.,* 92, 30–37.

Metzger, R., F.J. Sartoris, M. Langenbuch, and H.O. Portner. 2007. Influence of elevated CO_2 concentrations on thermal tolerance of the edible crab *Cancer pagurus, J. Thermal Biol.,* 32, 144–151.

Michaelidis, B., C. Ouzounis, A. Paleras, and H.O. Portner. 2005. Effects of long-term moderate hypercapnia on acid-base balance and growth rate in marine mussels *Mytilus galloprovincialis, Mar. Ecol. Prog. Ser.,* 293, 109–118.

Michaelidis, B., A. Spring, and H.O. Portner. 2007. Effects of long-term acclimation to environmental hypercapnia on extra cellular acid-base status and metabolic capacity in the Mediterranean fish, *Sparus aurata, Mar. Biol.,* 150, 1417–1429.

Mickel, T.J. and J.J. Childress. 1978. The effect of pH on oxygen consumption and activity in the bathypelagic mysid *Gnathophausia ingens, Biol. Bull.,* 154, 138–147.

Miles, H., S. Widdicombe, J.I. Spicer, and J. Hall-Spencer. 2007. Effects of anthropogenic seawater acidification on acid-base balance in the sea urchin *Psammechinus miliaris, Mar. Pollut. Bull.,* 54, 89–96.

Miller, A.W., A.C. Reynolds, C. Sobrino, and G.R. Riedel. 2009. Shellfish face uncertain future in high CO_2 world: influence of acidification on oyster larvae calcification and growth in estuaries, *PloS ONE,* 4(5): e5661. doi:1371/journal. pone.0005661.

Moran, D. and J.G. Stottrup. 2010. The effect of carbon dioxide on growth of juvenile Atlantic cod *Gadus morhua* L., *Aquat. Toxicol.,* doi:10.1016/j. aquatox.2010.12.014.

Moran, D., R.M.G. Wells, and S.J. Pether. 2008. Low stress response exhibited by juvenile yellowtail kingfish (*Seriola lalandi* Valenciennes) exposed to hypercapnic conditions associated with transportation, *Aquacult. Res.,* 39, 1399–1407.

Morita, M., R. Suwa, A. Iguchi, M. Nakamura, K. Shimada, K. Sakai, and A. Suzuki. 2010. Ocean acidification reduces sperm flagellar motility in broadcast spawning reef invertebrates, *Zygote,* 18, 103–107.

Munday, P.L., D.L. Dixson, J.M. Donelson, G.P. Jones, M.S. Pratchett, G.V. Devitsina, and K.B. Doving. 2009. Ocean acidification impairs olfactory discrimination and homing ability of a marine fish, *Proc. Natl. Acad. Sci. USA,* 106, 1848–1852.

Munday, P.L., D.L. Dixson, M.I. McCormick, M. Meekan, M.C.O. Ferrari, and DP. Chivers. 2010. Replenishment of fish populations is threatened by ocean acidification, *Proc. Natl. Acad. Sci. USA,* 107, 12930–12934.

Munday, P.L., J.M. Donelson, D.L. Dixson, and J.G.K. Endo. 2009a. Effects of ocean acidification on the early life history of a tropical marine fish, *Proc. Roy. Soc.,* 276B, 3275–3283.

O'Donnell, M.J., L.M Hammond, and G.E. Hofmann. 2009. Predicted impact of ocean acidification on a marine invertebrate: elevated CO_2 alters response to thermal stress in sea urchin larvae, *Mar. Biol.*, 156, 439–446.

O'Donnell, M.J., A.E. Todgham, M.A. Sewell, L.M. Hammond, K. Ruggiero, N.A. Fangue, M.L. Zippay, and G.E. Hoffmann. 2010. Ocean acidification alters skeletogenesis and gene expression in larval sea urchins, *Mar. Ecol. Prog. Ser.*, 398, 157–171.

Ohde, S. and M.M.M. Hossain. 2004. Effect of $CaCO_3$ (aragonite) saturation state of seawater on calcification of *Porites* coral, *Geochem. J.*, 138, 613–621.

Omori, M., C.P. Norman, and T. Ikeda. 1998. Oceanic disposal of CO_2: potential effects on deep-sea plankton and micronekton. A review, *Plankton Biol. Ecol.*, 45, 87–99.

Pagano, G., M. Cipollaro, G. Corsale, A. Esposito, E. Ragucci, and G.G. Giordano. 1985. pH-induced changes in mitotic and developmental patterns in sea urchin embryogenesis. I. Exposure of embryos, *Teratogen., Carcinogen. Mutagen.*, 5, 101–112.

Palacios, S.L. and R.C. Zimmerman. 2007. Response of eelgrass *Zostera marina* to CO_2 enrichment: possible impacts of climate change and potential for remediation of coastal habitats, *Mar. Ecol. Prog Ser.*, 344, 1–13.

Pane, E.F. and J.P. Barry. 2007. Extracellular acid-base regulation during short-term hypercapnia is effective in shallow-water crab, but ineffective in a deep-sea crab, *Mar. Ecol. Prog. Ser.*, 334, 1–9.

Park, D., Y.S. Hun, C.K. Ahn, and J.M. Park. 2007. Kinetics of the reduction of hexavalent chromium with the brown seaweed *Ecklonia* biomass, *Chemosphere*, 66, 939–946.

Parker, L.M., P.M. Ross, and W. O'Connor. 2009. The effect of ocean acidification and temperature on the fertilization and embryonic development of the Sydney rock oyster *Saccostrea glomerata* (Gould 1850), *Global Change Biol.*, 15, 2123–2136.

Pond, D.W., R.P. Harris, and C. Brownlee. 1995. A microinjection technique using a pH-sensitive dye to determine the gut pH of *Calanus helgolandicus*, *Mar. Biol.*, 123, 75–79.

Porter, J.W. and O.W. Meier. 1992. Quantification of loss and change in Floridian reef coral populations, *Amer. Zool.*, 32, 635–640,

Portner, H.O. 2008. Ecosystem effects of ocean acidification in times of ocean warming: a physiologist's view, *Mar. Ecol. Prog. Ser.*, 373, 203–217.

Portner, H.O., M. Langenbuch, and A Reipschlager. 2004. Biological impact of elevated ocean CO_2 concentrations: lessons from animal physiology and earth history, *J. Ocean.*, 60, 705–718.

Raven, J., K. Caldeira, H. Elderfield, O, Hoegh-Guldberg, P. Liss, U. Riebesell, J. Shepard, C. Turley, and A. Watson. 2005. Ocean acidification due to increasing atmospheric carbon dioxide, *Policy doc. 12/05, The Royal Society, 6–9 Carlton House Terrace, London SW15AG*, 57 pp.

Reynaud, S., N. Leclercq, S. Romaine-Lioud, C. Ferrier-Pages, J. Laubert, and J.P. Gattuso. 2003. Interacting effects of CO_2 partial pressure and temperature on photosynthesis and calcification in a scleractinian coral, *Global Change Biology*, 9, 1660–1668.

Riebesell, U. 2004. Effects of CO_2 enrichment on marine phytoplankton, *J. Ocean*, 60, 719–729.

Riebesell, U., K.G. Schulz, R.G.J. Bellerby, M. Botros, P. Fritsche, M. Meyerhofer, C. Neill, G. Nondal, A. Oschlies, J. Wohlers, and E. Zollner. 2007. Enhanced biological carbon consumption in a high CO_2 Ocean, *Nature*, 450, 545–548.

Riebesell, U., D.A. Wolf-Gladrow, and V. Smetacek. 1993. Carbon dioxide limitation of marine phytoplankton growth rates, *Nature*, 361, 249–251.

Riebesell, U., I. Zondervan, B. Rost, P.D. Tortell, R.E. Zeebe, and F.M.M. Morel. 2000. Reduced calcification of marine plankton in response to increased atmospheric CO_2 , *Nature*, 407, 364–367.

Ries, J.B., A.L. Cohen, and D.C. McCorkle. 2009. Marine calcifiers exhibit mixed responses to CO_2-induced ocean acidification, *Geology*, 37, 1131–1134.

Rijnsdorp, A.D., M.A. Peck, G.H. Engelhard, C. Mollmann, and J.K. Pinnegar. 2009. Resolving the effect of climate change on fish populations, *J. Mar. Sci.*, 66, 1570–1583.

Ringwood, A.H. and C.J. Keppler. 2002. Water quality variation and clam growth: is pH really a non-issue in estuaries, *Estuaries*, 25, 901–907.

Ruttimann, J. 2006. Oceanography: sick seas, *Nature*, 442, 978–980.

Scheuhammer, A.M. 1991. Effects of acidification on the availability of toxic metals and calcium to wild birds and mammals, *Environ. Pollut.*, 71, 329–375.

Schippers, P., M. Lurling, and M. Scheffer. 2004. Increase of atmospheric CO_2 promotes phytoplankton productivity, *Ecol. Lett.*, 7, 446–451.

Sciandra, A., J. Harlay, D. Lefevre, R. Lemee, P. Rimmelin, M. Denis, and J.P. Gattuso. 2003. Response of coccolithophorid *Emiliana huxleyi* to elevated partial pressure of CO_2 under nitrogen limitation, *Mar. Ecol. Prog. Ser.*, 261, 111–122.

Shearer, T.L. and M.A. Coffroth. 2006. Genetic identification of Caribbean scleractinian coral recruits at the Flower Garden Banks and the Florida Keys, *Mar. Ecol. Prog. Ser.*, 306, 133–142.

Shiber, J. and E. Washburn. 1978. Lead, mercury, and certain nutrient elements in *Ulva lactuca* (Linnaeus) from Ras Beirut, Lebanon, *Hydrobiologia*, 61, 187–192.

Shirayama, Y. and H. Thornton. 2005. Effect of increased atmospheric CO_2 on shallow water marine benthos, *J. Geophys. Res*, 110, C09S08, doi:10.1029/2004JC002618.

Siikavuoplo, S.I., A. Mortensen, T. Dale, and A. Foss. 2007. Effects of carbon dioxide exposure on feed intake and gonad growth in green sea urchin, *Strongylocentrotus droebachiensis*, *Aquaculture*, 266, 97–101.

Spicer, J.I., A. Raffo, and S. Widdicombe. 2007. Influence of CO_2-related seawater acidification on extracellular acid-base balance in the velvet swimming crab, *Necora puber*, *Mar. Biol.*, 151, 1117–1125.

Spicer, J.I., A.C. Taylor, and A.D. Hill. 1988. Acid-base status in the sea urchins *Psammechinus miliaris* and *Echinus esculentus* (Echinodermata:Echinoidea) during emersion, *Mar. Biol.*, 99, 27–53.

Stanley, G.D., Jr. 2007. Ocean acidification and scleractinian corals, *Science*, 317, 1032.

Stone, R. 2007. A world without corals?, *Science*, 316 (5825), 678–681.

Sunda, W., and R.R. Guillard. 1976. The relationship between cupric ion activity and the toxicity of copper to phytoplankton, *J. Mar. Res.*, 34, 511–529.

Sunda, W.G., P.A. Tester, and S.A. Huntsman. 1990. Toxicity of trace metals to *Acartia tonsa* in the Elizabeth River and southern Chesapeake Bay, *Estuar. Coastal Shelf Sci.*, 30, 207–221.

Talmage, S.C. and C.J. Gobler. 2009. The effects of elevated carbon dioxide concentrations on the metamorphosis, size, and survival of larval hard clams (*Mercenaria mercenaria*), bay scallops (*Argopecten irradians*), and eastern oysters (*Crassostrea virginica*), *Limnol. Ocean*, 54, 2072–2080.

Talmage, S.C. and C.J. Gobler. 2010. Effects of past, present, and future ocean carbon dioxide concentrations on the growth and survival of larval shellfish, *Proc. Natl. Acad. Sci. USA*, 107, 17246–17251.

Tanner, C.A., L.E. Burnett, and K.G. Burnett. 2006. The effects of hypoxia and pH on phenoloxidase activity in the Atlantic blue crab, *Callinectes sapidus*, *Comp. Biochem. Physiol.*, 144A, 218–223.

Thistle, D., K.R. Carman, L. Sedlacek, P.G. Brewer, J.W. Fleeger, and J.P. Barry. 2005. Deep-ocean, sediment-dwelling animals are sensitive to sequestered carbon dioxide, *Mar. Ecol. Prog. Ser.*, 289, 1–4.

Todgham, A.E. and G.E. Hofmann. 2009. Transcriptomic response of sea urchin larvae *Strongylocentrotus purpuratus* to CO_2-driven seawater acidification, *J. Exp. Biol.*, 212, 2579–2594.

Turley, C.M., J.M. Roberts, and J.M. Guinotte. 2007. Corals in deep-water: will the unseen hand of ocean acidification destroy cold-water ecosystems?, *Coral Reefs*, 26, 445–448.

UNEP (United Nations Environment Programme). 2010. Environmental consequences of ocean acidification: a threat to food security, *Available from UNEP*, P.O. Box 39552, Nairobi 00100, Kenya, 12 pp.

USDC (United States Department of Commerce), National Oceanic and Atmospheric Administration (NOAA). 2008. Ocean acidification, *State of the Science Fact Sheet*, 2 pp.

Venn, A.A., E. Tambutte, S. Lotto, D. Zoccola, D. Allemand, and S. Tambutte. 2009. Imaging intracellular pH in a reef coral and symbiotic anemone, *Proc. Natl . Acad. Sci. USA*, 106, 16574–16579.

Vetter, E.W. and C.R. Smith. 2005. Insights into the ecological effects of deep ocean CO_2 enrichment: the impacts of natural CO_2 venting at Loihi seamount on deep sea scavengers, *J. Geophys. Res.*, 110, C09S13, doi:10.1029/2004JC002617.

Waldbusser, G.G., E.P. Voigt, H. Bergschneider, M.A. Green, and R.I.E. Newell. 2010. Biocalcification in the Eastern oyster (*Crassostrea virginica*) in relation to long-term trends in Chesapeake Bay pH, *Estuaries and Coasts*, doi:10.1007/s12237-010-9307-0.

Walther, K., F.J. Sartoris, C. Bock, and H.O. Portner. 2009. Impact of anthropogenic ocean acidification on thermal tolerance of the spider crab *Hyas araneus*, *Biogeosciences*, 6, 2207–2215.

Watanabe, Y., A. Yamaguchi, H. Ishida, T. Harimoto, S. Suzuki, Y. Sekido, Y. Shirayama, M.M Takahashi, and J. Ishizaka. 2006. Lethality of increasing

CO$_2$ levels on deep-sea copepods in the western North Atlantic. *J. Ocean*, 62, 185–196.

Welch, C. 2009. Oysters in deep trouble: is Pacific Ocean's chemistry killing sea life?, *Seattle Times* (newspaper), June 14, 2009.

Wickins, J.F. 1984. The effect of hypercapnic sea water on growth and mineralization in penaied prawns, *Aquaculture*, 41, 37–48.

Widdicombe, S., S.L. Dashfield, C.L. McNeill, H.R. Needham, A. Beesley, A. McEvoy, S. Oxnevad, K.R. Clarke, and J.A. Berge. 2009. Effects of CO$_2$ induced seawater acidification on infaunal diversity and sediment nutrient fluxes, *Mar. Ecol. Prog. Ser.*, 379, 59–75.

Widdicombe, S. and H.R. Needham. 2007. Impact of CO$_2$-induced seawater acidification on the burrowing activity of *Nereis virens* and sediment nutrient flux, *Mar. Ecol. Prog. Ser.*, 341, 111–122.

Wood, H.L., J.I. Spicer, and S. Widdicombe. 2008. Ocean acidification may increase calcification rates, but at a cost, *Proc. Royal Soc.*, 275B, 1767–1773.

Wood, H.L., S. Widdicombe, and J.I. Spicer. 2009. The influence of hypercapnia and the infaunal brittlestar *Amphiura filiformis* on sediment nutrient flux–will ocean acidification affect nutrient exchange?, *Biogeosciences*, 6, 2015–2024.

Wootton, J.T., C.A. Pfister, and J.D. Forester. 2008. Dynamic patterns and ecological impacts of declining ocean pH in a high-resolution multi-year dataset, *Proc. Natl Acad. Sci. USA*, 105, 18848–18853.

Wyatt, N.J., V. Kitidis, E. Malcolm, S. Woodward, A.P. Rees, S. Widdicombe, and M. Lohan. 2010. Effects of high CO$_2$ on the fixed nitrogen inventory of the Western English Channel, *J. Plankton Res.*, 32, 631–641.

Yamada, Y. and T. Ikeda. 1999. Acute toxicity of lowered pH to some oceanic zooplankton, *Plankton Biol. Ecol.*, 46, 62–67.

Zaroogian, G.E., G. Morrison, and J.F. Heltshe. 1979. *Crassostrea virginica* as an indicator of lead pollution, *Mar. Biol.*, 52, 189–196.

Zhao, Y. and S. Sun. 2006. Effects of salinity, temperature, and pH on the survival of the nemertean *Procephalothrix simulus* Iwata, 1952, *J. Exp. Mar. Biol. Ecol.*, 328, 168–176.

Zimmerman, R.C., D.G. Kohrs, D.L. Steller, and R.S. Alberte. 1997. Impacts of CO$_2$ enrichment on productivity and light requirements of eelgrass, *Plant Physiol.*, 115, 599–607.

Zippay, M.L. and G.E. Hofmann. 2010. Effect of pH on gene expression and thermal tolerance of early life history stages of red abalone (*Haliotis rufescens*), *J. Shellfish Res.*, 29, 429–439.

Zondervan, I., B. Rost, and U. Riebesell. 2002. Effect of CO$_2$ concentration on the PIC/POC ratio in the coccolithophore *Emiliana huxleyi* grown under light-limiting conditions and different daylengths, *J .Exp. Mar. Biol. Ecol.*, 272, 55–70.

CHAPTER 6

Field Studies

6.1 General

Model simulations based on ocean circulation of atmospheric CO_2 emissions suggest that stabilization of atmospheric CO_2 at 450 ppm produces both calcite and aragonite undersaturation in most of the deep ocean (Caldeira and Wickett 2005). Although the physical and chemical basis for ocean acidification is well understood, there exist few field data of sufficient duration, resolution, and accuracy to document the acidification rate and to elucidate the factors governing its variability (Dore et al. 2009).

To speculate on the future change of CO_2-induced climate changes over several centuries, three 500-year integrations of a coupled ocean-atmosphere model were performed (Manabe and Stouffer 1994). In one integration the concentration of atmospheric CO_2 remained unchanged. In another, CO_2 concentration increased by 1% per year (compounded) until it reaches four times the initial value at the 140th year and remains unchanged thereafter. In the third model, atmospheric CO_2 also increases by 1% per year until it reaches twice the initial value at the 70th year and remains unchanged thereafter. In the CO_2-quadrupling integration, thermohaline circulation in the North Atlantic Ocean and the Weddell and Ross Seas vanish during the first 250 years, and is attributed mainly to the capping of the model oceans by relatively fresh water in high latitudes. In the CO_2-doubling integration, thermohaline circulation weakens by a factor of more than 2 during the first 150 years but almost recovers by year 500, aided by the gradual increase in surface salinity. During the 500-year period of the doubling and quadrupling models, the global mean surface air temperature increased by about 3.5°C and 7.0°C, respectively. The rise of sea level due to the thermal expansion of seawater is predicted to be about 1.0 and 1.8 m, respectively, and could be much larger if the contribution of meltwater from continental ice sheets were included (Manabe and Stouffer 1994).

Effects—both actual and projected—of CO_2 discharged into the environment as a result of human activities are documented below for selected marine bodies of water.

6.2 Arctic Ocean

Data based on four different expeditions show a lysocline (= the depth at which $CaCO_3$ solubility increases substantially) for aragonite around 3,500 m and for calcite most of the sea floor above the lysocline (Jutterstrom and Anderson 2005). Conditions seem favorable for $CaCO_3$ preservation despite the depths and cold temperature and the low production of $CaCO_3$. This is probably a result of low sedimentation and respiration rates of soft organic materials. However, a slight increase in the sedimentation and decay of soft organic material caused by decreasing sea ice cover in the summer can make the deep water corrosive enough to induce severe shell dissolution (Jutterstrom and Anderson 2005). Surface waters of the Arctic Ocean are vulnerable to becoming undersaturated with respect to aragonite during the 21st century (Orr et al. 2006). Decreases in Arctic carbonate ion concentrations are likely to affect many calcifying organisms including aragonitic cold-water corals and shelled pteropods, calcitic coccolithophores and foraminiferans, and high-magnesium calcite-producing coralline red algae and echinoderms. Limited data are available on the response of Arctic calcifiers to decreased carbonate saturation state, but evidence from lower latitudes suggests that at least some Arctic calcifiers will suffer reduced calcification, potentially affecting their competitiveness and survival. The added pressure of ocean acidification could reduce biodiversity and alter food-web structure of planktonic and benthic Arctic ecosystems (Orr et al. 2006).

By the year 2070, the largest simulated pH changes worldwide will occur in Arctic surface waters where hydrogen ion concentration is projected to increase by up to 185%, with a resultant pH decrease of

0.45 units (Steinacher et al. 2009). Projected climate change amplifies the decrease in Arctic surface mean aragonite saturation rate and pH by more than 20%, mainly due to freshening and increased carbon uptake in response to sea ice retreat. Aragonite undersaturation in Arctic surface waters is projected to occur locally within a decade and to become more widespread as atmospheric CO_2 concentrations increase above 450 ppm (Steinacher et al. 2009).

6.3 Arabian Sea

An estimate of the carbon budget for the Arabian Sea (0–25° N, 50–80° E) indicates that the annual carbon flux into and out of the Arabian Sea is 446 trillion grams (Tg) and 530 Tg, respectively (Somasundar et al. 1990). The 84 Tg C deficit is attributed to CO_2 loss into the atmosphere and to northward movement of bottom waters. Fluxes from rivers did not contribute significantly to the carbon influx, nor was burial of carbon through sedimentation processes. Authors' model indicates that the Arabian Sea is a carbon source for the Persian Gulf and the Red Sea. Based on the standing crop and net outfluxes, the estimated residence time for carbon in the Arabian Sea is 944 years (Somasundar et al. 1990).

6.4 Atlantic Ocean

Surface concentrations of anthropogenic CO_2 were highest in the tropical and subtropical regions and decreased toward the high latitudes (Gruber 1998). Highest specific inventories (inventory per m^2) of anthropogenic CO_2 occur in the subtropical convergence zones. Large differences are evident between the North and South Atlantic high latitudes: in the North Atlantic, anthropogenic CO_2 has invaded deeply into the interior and has reached the bottom north of 50° N. However, waters south of 50° S contain relatively little anthropogenic CO_2, and hence specific inventories are low. An anthropogenic CO_2 inventory of 22 Gt C (1 gigaton [Gt] = 1 x 10^{12} kg) is estimated for the Atlantic Ocean north of the equator for 1982, and 18 Gt C is estimated for the Atlantic Ocean south of the equator for 1989 (Gruber 1998). Surveys of the Atlantic Ocean between 1990 and 1998 show that anthropogenic CO_2 inventories are highest in the subtropical regions at 20–40° S (Lee et al. 2003a).

However, anthropogenic CO_2 penetrates the deepest in high-latitude regions >40° N owing to the formation of deep water that feeds the Deep Western Boundary Current which transports anthropogenic CO_2 into the interior. The total anthropogenic CO_2 inventory was estimated at 28.4 Pg C (1 petagram [Pg] = 1 billion metric tons) in the North Atlantic Ocean (equator to 70° N), and 18.5 Pg C in the South Atlantic Ocean (equator to 70° S) (Lee et al. 2003a).

On the basis of whole foraminifera shell weights of *Globigerinoides sacculifer* and *Neogloboquadrina dutertrei* from Caribbean cores, carbonate ion concentrations at 1,800 m depth in the open Atlantic was about 14 *u*mol/kg greater during glacial times than during the Holocene (Broecker and Clark 2002). However, this reconstruction is considered tentative as unanswered questions remain regarding the conversion of shell weights to paleocarbonate ion concentrations (Bijma et al. 2002). Critical in this regard are the assumptions that shell-wall thicknesses were the same as current thickness at any given site during glacial time and that the 0.3 *u*g per *u*mol/kg relationship between shell weight and carbonate ion concentration is applicable over the entire deep water column (Broecker and Clark 2002).

6.4.1 North Atlantic Ocean

Baseline measurements of CO_2, pH, total inorganic carbon, total alkalinity, and other hydrologic parameters of surface and subsurface waters were obtained during the National Oceanic and Atmospheric Administration's Ocean Atmosphere Carbon Exchange Study expedition in the eastern North Atlantic during summer 1993 (Lee et al. 1997). Using a model developed by the Intergovernmental Panel on Climate Change, the North Atlantic thermohaline circulation weakens in all global warming scenarios and collapses at high levels of CO_2 (Joos et al. 1999). Projected changes in the marine carbon cycle have a modest impact on atmospheric CO_2 levels. Compared with the control, atmospheric CO_2 increased by 4% at year 2100 and 20% at year 2500, with the reduction in ocean carbon uptake explained by sea surface warming. The projected changes of the marine biological cycle compensate the reduction in downward mixing of anthropogenic carbon, except when the North Atlantic thermohaline circulation collapses (Joos et al. 1999).

Between 1983 and 2005, CO_2 and dissolved inorganic carbon concentrations of surface seawater near Bermuda increased annually at rates similar to that expected from ocean equilibration with increasing CO_2 in the atmosphere (Bates 2007). In addition, seawater pH, CO_3^{2-} ion concentrations and $CaCO_3$ saturation states also decreased over time. Overall, the region was an oceanic sink for CO_2. Interannual variability of summertime and autumn air-sea CO_2 flux rates were significantly influenced by season, by wind events (such as hurricanes), and by the North Atlantic Oscillation (Bates 2007). Between June 1994 and August 1995, a fully automated instrument on a merchant ship measured the partial pressure of carbon dioxide in the ocean surface and overlying atmosphere (Cooper et al. 1998). The ship traveled between the U.K. and the Caribbean, with each voyage being 5 weeks duration. The large scale seasonal changes in a broad region of the mid-Atlantic were consistent with seasonal temperature fluctuations. However, to the north and east of the region studied variations were more extreme due to deep winter mixing and spring-summer biological activity. Variations were most extreme in coastal and shelf waters where the effects of phytoplankton activity were more obvious than in the open ocean. The feasibility of unattended data acquisition in a cost-effective method is extolled (Cooper et al. 1998). Substantial variability in the CO_2 uptake by the North Atlantic on time scales of a few years through automated instrumentation aboard volunteer observing ships is documented (Watson et al. 2009). This approach offers the prospect of accurately monitoring the changing ocean CO_2 sink for those ocean basins that are well covered by shipping routes (Watson et al. 2009). One study compared sea surface CO_2 levels between 1994–1995 and 2002–2005 and concluded that there was more than a 50% decline, mostly due to increasing stratification and the carbon content of surface waters (Schuster and Watson 2007).

The North Atlantic spring bloom is one of the largest biological events in the ocean, and is characterized by dominance transitions from siliceous diatoms to calcareous coccolithophores algal groups (Feng et al. 2009). A shipboard continuous culture experiment was conducted in June 2005 during this transition period to evaluate 4 treatments: 1/ 12°C + 390 mg/L CO_2 (ambient control); 2/ 12°C + 690 mg/L CO_2 (high pCO_2); 3/ 16°C + 390 mg/L CO_2 (high temperature); and 4/ 16°C+690 CO_2 (greenhouse). Nutrient availability in all treatments reproduced low silicate conditions typical of his late stage of the bloom. Both elevated CO_2 and temperature resulted in changes in phytoplankton community structure. Increased temperature promoted whole community photosynthesis and particulate organic carbon production rates per unit chlorophyll-*a*. Despite much higher coccolithophore abundance in the greenhouse treatment, calcification (particulate inorganic carbon production) was significantly decreased by the combination of increased temperature and CO_2. It is speculated that future trends during the bloom could include reduced export of calcium carbonate relative to particulate organic carbon, thus providing a potential negative feedback to atmospheric CO_2 concentration. Another trend with potential climate feedback is decreased community biogenic silica to particulate organic carbon ratio at higher temperature. More research seems needed on the interaction of elevated temperature/ CO_2 over decades-long time scales (Feng et al. 2009).

6.4.2 Subtropical North Atlantic Ocean

Since 1983, a long-term time series of seawater CO_2 and pH in the subtropical North Atlantic Ocean near Bermuda were used to evaluate the influence of acidic deposition on ocean acidity (Bates and Peters 2007). Acidifying deposition had negligible influence on seawater CO_2 chemistry of the Bermuda coral reef, with no evident impact on hard coral calcification. Moreover, wet deposition of nitrate had negligible effect (i.e., <0.5%) on annual rates of oceanic primary production in the North Atlantic Ocean and on the Bermuda coral reef system (Bates and Peters 2007).

Organic carbon fluxes to the deep ocean may be enhanced with ballast material such as calcite and opal (Poulton et al. 2006, 2006a). Simultaneous measurements of the open ocean production of calcite (calcification), opal (silicification), and organic carbon (photosynthesis) at 14 stations between 42°S and 49°N in the Atlantic Ocean were made. Mineralizing phytoplankton represented 5% to 20% of organic carbon fixation, with similar contributions from both diatoms and coccolithophores. Average turnover times for calcite and phytoplankton carbon are about 3 days, suggesting a relatively labile nature, By comparison, average turnover times for opal and particulate organic carbon are about 10 days. Rapid turnover of calcite suggests an important role for the plankton community in removing calcite from the upper ocean. A comparison of surface production rates to sediment trap data confirms that about 70% of calcite is dissolved in the upper 2–3 km, and only a small proportion (<2%) of total organic carbon reaches the deep ocean.

Between 1988 and 1998, surface seawater total CO_2 increased at a rate of 2.2 $umol/kg$ annually in the Sargasso Sea, located in the North Atlantic subtropical gyre (Bates 2001). During the same period, the partial pressure of CO_2 (pCO_2) of seawater increased at a rate similar to the rate of increase in atmospheric pCO_2. Significant correlations existed between anomalies of temperature, salinity, primary production, CO_2, alkalinity, and various current oscillations–and these may account, in part, for the anomalies documented in interannual trends (Bates 2001).

6.5 Australia

Skeletal density, linear extension, and calcification rate in massive *Porites* spp. coral colonies from two nearshore regions of the northern Great Barrier Reef, Australia, were examined between 1988 and 2003 (Cooper et al. 2008). Over the 16-year study period, calcification rates in *Porites* declined by approximately 21%, linear extension by about 16%, skeletal density by 0.36%, and annual extension by 1.02%. These findings were consistent with results of studies on the synergistic effects of elevated seawater temperatures and increasing CO_2 on coral calcification. More data on seawater chemistry of the Great Barrier Reef are needed to better understand the links between environmental change and effects on coral growth (Cooper et al. 2008). In another study, a total of 328 *Porites* coral colonies from 69 reefs of the Great Barrier Reef (GBR) ranging from coastal to oceanic locations and covering most of the >2,000 km length of the GBR were investigated between 1990 and 2005 (De'ath et al. 2009). There was a 14.2% decline in calcification rate between 1990 and 2005, mainly because linear growth declined by 13.3%. This severe and sudden decline in calcification was unprecedented in at least the past 400 years. The causes of the decline remain unknown, but may be associated with a declining saturation state of seawater aragonite, thus effectively diminishing the ability of GBR corals to deposit calcium carbonate (De'ath et al. 2009). Calcification by four species of crustose coralline algae on the northern GBR was estimated by combining measurements of oxygen, pH, and total alkalinity with equations describing the seawater carbonate equilibrium (Chisholm 2000). A significant fraction of all samples exhibited $CaCO_3$ dissolution in the dark, suggesting that erosive processes remove much of the $CaCO_3$ deposited annually by crustose coralline algae on GBR windward reef margins (Chisholm 2000).

6.6 Baltic Sea

Summer blooms of filamentous cyanobacteria, mainly *Aphanizomenon* sp. and *Nodularia spumigena,* are characteristic in the Baltic Sea (Ploug 2008). Photosynthesis is substantial within these cyanobacterial layers accumulating at the air-water interface Author suggests that the close association of autotrophic and heterotrophic organisms and processes creates a pH microenvironment that is favorable for iron uptake for the cyanobacteria, which, in turn, may release surplus nitrogen to the heterotrophic community. The pH varied from 7.4 in darkness to 9.0 in saturating light intensities (Ploug 2008).

6.7 Belgian Coastal Areas

There is high variability of CO_2 partial pressure in surface seawater along the coast of Belgium—a part of the North Sea (see 6.19)—between January 1995 and June 1996, at both daily and seasonal time scales (Borges and Frankignoulle 1999). Daily variations depend mainly on the tides, and seasonal variations in riverine input and phytoplankton biomass. During winter, the plume of the Scheldt River is oversaturated in CO_2 with respect to the atmosphere. During spring and summer, phytoplankton blooms occur in the lower Scheldt estuary and may lead to undersaturation of CO_2, although degradation of phytoplankton induces oversaturation. On an annual basis, the Scheldt plume is a net source of CO_2 to the atmosphere (Borges and Frankignoulle 1999).

6.8 Bering Sea and Environs

Biogenic particle fluxes in the Bering Sea and the central subarctic Pacific Ocean were measured with sediment traps 600 m from the bottom over the 5-year period 1990–1995 (Takahashi et al. 2000). The subarctic pelagic station (49° N, 174° W, water depth 5,406 m) was simultaneously studied along with the marginal station (53.5° N, 177° W; water depth 3,788 m) located in the Aleutian Basin of the Bering Sea. The ratios of organic carbon/inorganic carbon were usually greater than 1.0 at both stations, suggesting that preferentially greater organic carbon from cytoplasm than skeletal inorganic carbon was exported to the surface layers. This process, known as the biological pump, leads to a carbon sink which effectively lowers pCO_2 in the surface layers and then allows a net flux of CO_2 into the surface layer. The efficiency of the biological pump is greater in the Bering Sea than at the open-ocean station (Takahashi et al. 2000).

The Bering Sea is now experiencing changes in sea surface temperature, unprecedented algal blooms, and alterations to trophic level dynamics (Hare et al. 2007). Phytoplankton communities from two Bering Sea regimes that were incubated under conditions of elevated sea surface temperature and CO_2 levels predicted for the year 2100, showed increased photosynthetic rates up to 3.5 times and community composition shifted towards nanophytoplankton from diatoms. These changes were driven mainly by elevated temperature with secondary effects from increased CO_2. If these results are indicative of future climate responses, community shifts toward nanophytoplankton dominance could reduce the ability of the Bering Sea to maintain the diatom-based food webs that currently support a productive fishery (Hare et al. 2007).

6.9 Bermuda

Rising atmospheric CO_2 and ocean acidification originating from human activities could result in increased dissolution of metastable carbonate minerals in shallow water marine sediments (Andersson et al. 2007). *In situ* dissolution of carbonate sedimentary particles at Devil's Hole, Bermuda, was observed during summer when thermally driven density stratification restricted mixing between the surface and bottom waters, and microbial decomposition of organic matter in the subthermocline layer produced CO_2 levels similar to levels anticipated by the the year 2100. Mg-calcite minerals are preferentially subjected to dissolution under conditions of elevated CO_2, with a projected yearly range of 175 to 701 g $CaCO_3$/ m^2, the latter rate approximating 50% of the current annual average global reef calcification rate of 1,500 g $CaCO_3/m^2$. Authors aver that a reduction in marine calcification of 40% by the year 2100, or 90% by the year 2300, as a result of surface ocean acidification and increasing atmospheric CO_2 from burning of fossil fuels—in combination with high rates of carbonate dissolution and reduced rates of calcification—would result in a net loss in carbonate material of coral reefs and other carbonate sediment environments (Andersson et al. 2007).

6.10 Borneo

Shell dissolution in the gastropod whelk, *Thais gradata*, was observed in the Sungai Brunei estuary (Borneo), a typical South East Asian estuary with acidic waters (Marshall et al. 2008). Shell weight, shell length, and topographical shell features were determined for populations of *T. gradata* distributed along a pH gradient between 5.78 and 8.30 in 2005–2007. Decreasing shell length and increasing shell erosion were significantly correlated with declining pH (Marshall et al. 2008).

6.11 Caribbean Region

The global oceans serve as the largest sustained natural sink for increasing atmospheric carbon dioxide. As this CO_2 is absorbed by seawater it lowers the pH and decreases the carbonate mineral saturation state (Ω), which is key in the calcification process for many species of calcifiers (Gledhill et al. 2008). Observations obtained from Volunteer Observing Ships, multiple geochemical surveys, and satellite remote sensing, has enabled construction of a model to derive estimates of sea-surface alkalinity and CO_2 partial pressure. Pairing of these two parameters have permitted characterization of the changes in sea-surface Ω that have transpired during the period 1996–2006 throughout the Greater Caribbean Region as a consequence of ocean acidification. Despite considerable spatial and temporal variability, a strong decrease in aragonite saturation state (Ω_{arg}) was observed at an approximate rate of –0.012 Ω_{arg} annually (Gledhill et al. 2008).

6.12 Greenland Sea

Coccolithophore fluxes were investigated by sediment traps at 300, 900, and 2,100 m depths in the central Greenland Sea from 1994 to 1995 (Andruleit 2000). The dominant species were *Coccolithus pelagicus* (>90%) and *Emiliana huxleyi* (<7%). Very low fluxes were recorded during most of the investigated time interval. The coccolithophore settling assemblages of the traps were strongly influenced by dissolution. The dissolution impact probably diminished diversity and sharpened seasonality in fluxes. Dissolution was prominent in near-surface waters of the East Greenland Current, but not in bottom waters. Maximum fluxes were similar to that of the nearby Norwegian Sea and higher than in the southern Greenland Sea. In contrast to other sites, resuspension was minor. Total annual coccolithophore fluxes differed significantly in the Norwegian-Greenland Seas reflecting the coccolithophore production (Andruleit 2000).

6.13 Gulf of Maine

The Gulf of Maine is a continental shelf sea off the northeastern coast of the United States with a maximum depth of 275 m and an area of 90,700 km². Basic information has been collected since 1998 on hydrography, nutrients, phytoplankton abundance, and carbon fixation in response to several years of extreme river discharge (Balch et al. 2008). It was concluded that the Gulf of Maine is a biological sink for organic and inorganic carbon. Calcification was significantly correlated to primary production. Total phytoplankton synthesis was 38.1 Tg C/year and total calcification was 0.55 Tg C/year yielding an overall ratio of calcification to photosynthesis of 1.44%. The focus on standing stocks and production of organic carbon and inorganic carbon are considered relevant to understanding the impact of ocean acidification in coastal pelagic ecosystems (Balch et al. 2008).

Photosynthesis and calcification were measured in a bloom of coccolithophores in the Gulf of Maine during 1989 (Balch et al. 1992). The total carbon fixed during the 1989 bloom amounted to about 0.5% of the total Gulf of Maine production for the entire year Observations showed physiological evidence of stratification between surface and deep populations within the surface mixed layer. Of the total organic and inorganic carbon buried in one year in the Gulf of Maine, about 25% is attributable to the calcite carbon derived from the annual coccolithophore bloom (Balch et al. 1992).

Measurements of inorganic (calcification) and organic (photosynthetic) carbon production were made in the Gulf of Maine during 1996 throughout the year (Graziano et al. 2000). In June, organic calcite production averaged 5% of total carbon production, or 26 mg C m^2 daily. In June, inorganic calcite production was >10% of total carbon production. The turnover time of calcite particles in the water column was about 12 days in June and 200 days in November when calcite standing stocks were high and calcification rates low. Yearly carbon production for the Gulf of Maine was estimated at 182.0 g m^2 organic carbon and 3.7 g m^2 inorganic carbon. If 1% of the organic carbon produced were buried in sediments, and 50% of the inorganic carbon, the result would be an approximately equal amount of each deposited in Gulf sediments. Inorganic carbon production by coccolithophores is an important contributor to Gulf sediments (Graziano et al. 2000).

6.14 Indian Ocean

Anthropogenic CO_2 inventory estimates for the Indian Ocean in 1995 indicate that the highest concentrations and deepest penetrations of anthropogenic carbon are associated with the Subtropical Convergence near 30° to 40° S and the lowest were observed south of 50° S (Sabine et al. 1999). Total anthropogenic CO_2 inventory north of 35° S was 13.6 Pg C (1 Pg = one billion tons); another 6.7 Pg C were stored in the Indian sector of the Southern Ocean giving a total Indian Ocean inventory of 20.3 Pg C for the year 1995 (Sabine et al. 1999). A different model, however, predicts an Indian Ocean sink north of 35° S that is only 0.61–0.68 times the results of the previous model while the Southern Ocean sink is nearly 2.6 times higher than the measurement-based estimate. These results clearly identify areas in the models that need further examination and provide a baseline for future studies (Sabine et al. 1999).

Anthropogenic CO_2 has increased the shallow dissolved inorganic carbon in the Indian Ocean by as much as 3% (Sabine et al. 2002, 2002a). Calcite saturation depths range from 2,900 to 3,900 m with the deepest saturation depths in the central Indian Ocean. Variations of aragonite saturation depth (200 m to 1,400 m) are similar to calcite, but the deepest saturations are in the southwestern Indian Ocean. The shallowest aragonite saturation depths are in the Bay of Bengal. However, in the northern Arabian Sea and Bay of Bengal the current aragonite saturations are 100 to 200 m shallower than in the 1790s. Estimates of carbonate dissolution rates range up to 0.083 $umol/kg$ annually in deep waters and 0.73 $umol/kg$ annually in upper waters, with a local maximum occurring in intermediate waters just below the aragonite saturation horizon. Dissolution is higher at 10–20°S than at lower latitudes (Sabine et al. 2002, 2002a).

6.15 Ischia Island, Italy

Effects of acidification on benthic ecosystems at shallow coastal sites where volcanic CO_2 vents lower the pH of the water column was investigated off the Mediterranean Island of Ischia (Hall-Spencer et al. 2008). Along gradients of normal pH (8.1–8.2) to lowered pH (7.8–7.9) to low pH (7.4–7.5), rocky shore communities of calcareous organisms shifted to communities absent reef-building corals and with significant reductions in sea urchin and coralline algal abundance. Seagrass production was highest at pH 7.6 where coralline algal biomass was significantly reduced and where gastropod shells were dissolving due to carbonate sub-saturation. Species populating the vent sites are resilient to naturally high concentrations of CO_2 and suggest that ocean acidity may benefit invasive non-native algal species (Hall-Spencer et al. 2008).

Colonies of the branched calcitic bryozoan *Myriapora truncata* were transplanted to normal (pH 8.1), high (pH 7.66), and extremely high CO_2 conditions (pH 7.43) at gas vents off Ischia Island in the Tyrrhenian Sea, Italy for up to 128 days at seawater temperatures of 19–24°C (Rodolfo-Metalpa et al. 2010). At pH 7.66, dead colonies dissolved, but survivors maintained the same net calcification rate as those growing at normal pH. Authors conclude that established colonies of *M. truncata* can withstand the levels of ocean acidification predicted in the next 200 years, possibly because the soft tissues protect the skeleton from an external decrease in pH; however, all colonies died at seawater temperatures of 25–28°C (Rodolfo-Metalpa et al. 2010).

6.16 Japan, Volcano Islands

Few marine organisms tolerate conditions where ocean pH falls significantly below the current value of about 8.1, and where calcite and aragonite saturation values are below unity. Nevertheless, dense clusters of the vent mussel *Bathymodiolus brevior* were observed in natural conditions of pH values between 5.36 and 7.29 on northwest Eifuku volcano, Mariana arc, where liquid carbon dioxide and hydrogen sulfide emerge in a hydrothermal setting (Tunnicliffe et al. 2009). Shell thickness and daily growth increments in shells from Eifuku were about half those recorded from mussels living in water with pH >7.8, suggesting that low pH may be implicated in metabolic impairment. Mussels of age 40 years were collected, with long survival attributed to intact shell covering due to absence of predatory crabs. The vulnerability of molluscs to predators is likely to increase in a future ocean with low pH, but, at present, mussels continue to precipitate shells in this low-pH environment (Tunnicliffe et al. 2009).

6.17 Labrador Sea

Seasonal cycles of CO_2 in surface waters of the central Labrador Sea (56.5° N, 52.6° W) were measured in 2004–2005 (Kortzinger et al. 2008). The region is characterized by a net CO_2 sink of 2.7 mol CO_2/m^2 that is mediated to a major extent by biological carbon drawdown during spring and summer. During winter, surface waters approach equilibrium with atmospheric CO_2. During 2005 a mixed layer carbon budget had a net community production of 4.0 mmol C/m^2, about one-third of which appears to undergo subsurface respiration in a depth range that is reventilated during the following winter. Wind speeds above 10 m/s are directly related to gas transfer coefficients for both CO_2 and O_2 (Kortzinger et al. 2008).

6.18 North American West Coast

A large section of the North American continental shelf is affected by ocean acidification (Feely et al. 2008). A hydrographic survey along the continental shelf of western North America from Central Canada to northern Mexico in May–June 2007 showed that seawater undersaturated with aragonite—a mineral form of $CaCO_3$—upwells into the continental shelf from 40–120 m to the surface. Although seasonal upwelling of aragonite-undersaturated waters onto the shelf occurs frequently, the ocean uptake of anthropogenic CO_2 has increased the areal extent of the affected area. These upwelling processes will expose coastal organisms living in the water column or at the sea floor to less aragonite-saturated waters, exacerbating the biological effects of oceanic acidification (Feely et al. 2008). These results were not predicted to occur in open ocean surface waters until 2050 (Orr et al. 2005).

6.19 North Sea

Carbon exchange fluxes with the North Atlantic Ocean dominates the gross carbon budget of the North Sea, a shelf sea on the NW European continental shelf (Thomas et al. 2005). The net carbon budget is mainly from carbon inputs of rivers, the Baltic Sea and the atmosphere. Based on existing data, the North Sea is a sink for organic carbon and more than 90% of the CO_2 taken up from the atmosphere is exported to the North Atlantic Ocean (Thomas et al. 2005).

Acidification and pH variability was projected for the southern North Sea based on a hydrodynamic model that incorporated photosynthesis and respiration, riverine boundary conditions, and atmospheric CO_2 concentrations (Blackford and Gilbert 2007). Annual pH changes varied from <0.2 in areas of low biological activity to >1.0 in areas influenced by riverine discharges. Acidification due to increased fluxes of atmospheric CO_2 was calculated and shown to exceed 0.1 pH units over the next 50 years and a total acidification of 0.5 pH units below pre-industrial levels at atmospheric CO_2 concentrations of 1,000 ppm. Inhibition of pelagic nitrification at decreasing pH is expected with concomitant effects on biogeochemistry, but the model is imperfect pending additional information on the complex interaction processes that govern biogeochemical cycles (Blackford and Gilbert 2007).

Abundance of jellyfish in the North Sea has increased since 1958, with frequency of jellyfish significantly greater since 1970, and with abundance positively correlated with decreasing pH (Attrill et al. 2007). All models produced under all climate change scenarios indicate a move toward a decreasing pH with rising CO_2 resulting in a significant increase in jellyfish abundance over the next 100 years (Attrill et al. 2007).

6.20 Pacific Ocean

The total anthropogenic CO_2 inventory over an area from 120° E to 70° W (excluding the South China Sea, the Yellow Sea, the Japan/ East Sea, and the Sea of Okhotsk) was 44.5 Pg C in 1994 (Sabine et al. 2002a). Approximately 28.0 Pg C was located in the southern hemisphere and 16.5 Pg C was located north of the equator. The deepest penetration of anthropogenic CO_2 was found at about 50° S. The shallowest penetration was found just north of the equator. Shallow anthropogenic CO_2 penetration was also observed in the high-latitude Southern Ocean. In the North Pacific Ocean a strong zonal gradient was observed in the anthropogenic CO_2 penetration depth with the deepest penetration in the western Pacific Ocean. The Pacific has the largest total inventory in all the southern latitudes. The lack of deep and bottom water formation in the North Pacific means that the North Pacific Ocean inventories are smaller than those of the North Atlantic Ocean (Sabine et al. 2002a).

6.20.1 Central North Pacific Ocean

Autotrophic carbon assimilation measurements were performed monthly between 1988 and 1992 in the North Pacific subtropical gyre (Letelier et al. 1996). Photosynthetic values ranged from 127 to 1,055 mg C/m² daily while the average carbon assimilation number (P^b), defined as carbon assimilation rate per unit chlorophyll-*a*, varied between 1.6 and 12.0 g chlorophyll-*a* per hour in the 0–45 m depth range. Consistently low P^b values were observed in the first 2 years of the study but increased during 1991–1992. This rise in P^b was not associated with an increase in chlorophyll *a* and occurred during a period of increased water column stability. Reduction in ATP and nitrate/nitrite concentrations in the upper euphotic zone suggests that nutrient injections due to mixing events were minor or absent after January 1991. Authors suggest that high P^b values during 1991–1992 are due to a reduction in the frequency of mixing events and/or an ecosystem shift from nitrogen to phosphorus limitation and conclude that nutrient dynamics within the euphotic zone of the North Pacific tropical gyre need to be understood in order to interpret changes in P^b and to predict carbon fluxes (Letelier et al. 1996).

In a 20-year study near Hawaii, time-series measurements of seawater pH and associated parameters were made (Dore et al. 2009). Authors found a significant long-term decreasing trend in surface pH that was indistinguishable from the rate of acidification expected for equilibration with the atmosphere. Superimposed upon this trend was a strong seasonal pH cycle driven by temperature, mixing, and net photosynthetic CO_2 assimilation. Substantial annual variability in surface pH was influenced by climate-induced fluctuations in upper ocean stability. Below the mixed layer, changes in acidification were enhanced within distinct subsurface strata. These zones are influenced by remote water mass formation and intrusion, biological carbon remineralization, or both. Physical and biogeochemical processes alter the acidification rate with depth and time and should be incorporated into future ocean pH predictive models (Dore et al. 2009).

6.20.2 Eastern North Pacific Ocean

Trophic coupling between a pelagic food supply and its use by the sediment community was conducted at 4,100-m depth in the eastern North Pacific between 1989 and 1998 (Smith et al. 2001). Particulate organic carbon and particulate total nitrogen flux declined significantly from 1989 to 1996 then increased in 1998. Sediment community oxygen consumption declined progressively for the first 7 years. Authors could not explain why there was a discrepancy between the supply of food to and the demand for food by the deep sea benthos, especially over the 8-year period during which episodic, seasonal, and interannual variability would be expected to average out. It is possible that the use of organic matter by the sediment community is fueled by sources not adequately quantified by measurement techniques and that major food events occur over a period of decades. There seems to be adequate supplies of food to sustain the sediment community during periods of low food supply and the decline in flux of sinking particulate organic matter reaching the deep-sea floor would lead to substantial changes in the structure and function of benthic communities as well as in the geochemistry of marine sediments (Smith et al. 2001).

6.20.3 Equatorial Pacific Ocean

Particle flux during all of 1992 to depths of 2.4–4.4 km, as measured by sediment traps every 17 days, varied between latitudes, season, particle composition, and upwelling, suggesting a settling particle residence time shorter than 17 days (Honjo et al. 1995). Calcite standing stock, calcification rate, concentrations of detached coccoliths and plated coccolithophore cells were measured in the equatorial Pacific along 140° W, between 12°N and 12° S latitude during August and September 1992 (Balch and Kilpatrick 1996). Variables known to modify these parameters include depth, upwelling, latitude, and pH. Calcification ranged between 3% and 12% of the total carbon fixed into particulate matter. Populations from the equator to 3° N at 60 m depth, and near the surface from the equator to 9° S were the most active calcite producers. Estimates of light scattering demonstrated the importance of upwelling of cold, clear, relatively particle-free water to the surface, followed by growth and calcite production with increasing temperature. Calcite particles were scarce in the top 1,000 m above the lysocline. Authors suggest that dissolution occurred where decomposition of reduced organic matter lowers the pH sufficiently to dissolve calcite (Balch and Kilpatrick 1996).

Downward transport of carbon by diel migrant mesozooplankton was measured in March and October of 1992 at the equator (Zhang and Dam 1997). Downward flux of carbon from the euphotic zone due to diel vertical migrators was 0.6 mmol C/m^2 daily in March and 1.1 mmol C m^2 daily in October. The migratory flux is strongly dependent on whether feeding occurs below the euphotic zone, on length of time migrators spend in deep waters, and on the mortality rate of migrators (Zhang and Dam 1997). A field incubation experiment demonstrated a substantial shift in the taxonomic composition of Equatorial Pacific phytoplankton assemblages exposed to 150 or 750 mg/L CO_2 during September 2000 (Tortell et al. 2002). The initial phytoplankton community preincubation was dominated by cryptophores (45% of the total) and prymnesiophytes (13% of total). By the end of the 11-day experiment, the phytoplankton community in all samples was dominated by diatoms and *Phaeocystis* spp. (a prymnesiophyte), but the relative abundance differed significantly between CO_2 regimens. Abundance of diatoms decreased by about 50% at low CO_2 relative to high CO_2, while the abundance of *Phaeocystis* increased by about 60% at low CO_2. This CO_2-dependent shift was associated with a major change in nutrient use, with higher ratios of nitrate:silicate and nitrate:phosphate consumption by phytoplankton in the low CO_2 treatment. Despite the changes in taxonomic composition and nutrient consumption ratios, total biomass and primary productivity did not differ significantly between the CO_2 treatments. These CO_2 levels could influence competition among phytoplankton taxa and affect oceanic nutrient cycling (Tortell et al. 2002).

6.20.4 Northeast Pacific Ocean

One of the world's highest abundances of glass sponges occurs in fjords and continental shelf waters of the northeast Pacific covering thousands of kilometers along the British Columbia and Alaska coastline (Yahel et al. 2007). Feeding and metabolism of two species, *Aphrocallistes vastus* and *Rhadocalyptus dawsoni*, was studied in a deep temperate fjord with a remotely operated submersible. Diet consisted mostly of bacteria and heterotrophic protists. The relatively scarce microbial cells were efficiently selected from a soup of clay and detritus particles where microorganisms accounted for 0.1% of the total ambient solids. This diet accounted for the entire total organic carbon uptake and ammonium excretion by both species, with no evidence for dissolved organic uptake. Similar results were obtained in laboratory experiments in which dissolved organic carbon was measured directly. Sponge filtering activity may significantly affect the deep microbial community and benthic-pelagic mass exchange in some northeast Pacific Ocean fjords (Yahel et al. 2007).

Early impressions of a vast deep ocean supplied with a constant rain of small organic particles from surface waters (Martin et al. 1987) is now replaced by a dynamic environment experiencing fluctuations in food supply on time scales of days to years (Smith et al. 2006). A 15-year study (1989–2004) to measure particulate organic carbon flux in the northeast Pacific as an estimate of food supply reaching depths of more than 4,000 m was conducted. Export flux from the euphotic zone was significantly correlated with particulate organic carbon, lagged earlier by 0–3 months (Smith et al. 2006). Seasonal and longer-term variations in particulate organic carbon flux were correlated to abyssal metazoan macrofauna abundance, community composition and structure, and carbon remineralization by using data from a 10-year study conducted at a northeast Pacific study site (34°50′ N, 123°00′ W, 4,100 m depth) (Ruhl et al. 2008). The majority of deep-sea communities depend on a particulate organic carbon food supply that sinks from photosynthetically active surface waters. Abyssal sediment communities and constituent metazoan fauna influence carbon and nutrient cycle processes through remineralization, bioturbation, and burial of the sunken material. Deep-sea macrofauna community structure can change with changes in surface ocean conditions and need to be considered in determining the long-term carbon storage capacity of the ocean (Ruhl et al. 2008). The carbonate carbon fluxes in the productive northern Pacific are larger in the northeast Pacific than in the northwest Pacific, reflecting a difference in the ecosystems (Tsunogai and Noriki 1991). This suggests that the eutrophication of marine environments may not necessarily act as a sink for the atmospheric carbon dioxide (Tsunogai and Noriki 1991).

6.20.5 Northwest Pacific Ocean

The fossil fuel CO_2 signal in seawater is strongly present in the upper 400 m, but estimates between models and observations vary by at least a factor of two (Brewer et al. 1997). In the North Pacific Ocean, the signal decays with depth, corresponding to the ventilation age of the water masses, and may be traced to about 400 m deep, consistent with deep winter time mixed layer formation in the northwest sector. However, insufficient data from the preindustrial ocean blurs final calculation. Refining the calculation will require increased knowledge of surface winter time total alkalinity values (Brewer et al. 1997).

Surface water CO_2 concentrations collected between 1970 and 2004 increased at a mean decadal rate of 12.0 mg/L in all but four areas in the vicinity of the Bering Sea and Okhotsk Sea where they decreased at a mean rate of 11.1 mg/L per decade (Takahashi et al. 2006). The mean rate of increase for the open ocean areas is indistinguishable from the mean atmospheric CO_2 increase rate of 15 ppm per decade, or 1.5 ppm yearly, suggesting that North Pacific Ocean surface waters reflect atmospheric CO_2 increase. However, the rate of increase varies geographically, reflecting differences in local oceanographic processes including lateral mixing of waters from marginal seas, upwelling of subsurface waters, and biological activities. The decrease observed in the southern Bering Sea and the peripheries of the Okhotsk Sea were attributed to the combined effects of intensified biological production and changes in lateral and vertical mixing in these areas (Takahashi et al. 2006).

A new research program on acidification effects is planned for the Bering Sea and North Pacific Ocean that will focus on commercially important fish and shellfish species, their prey (calcareous plankton) and shelter (corals) (Sigler et al. 2008). Species-specific studies of shellfish, calcareous plankton, corals, and fish will be conducted to understand effects on growth and survival, forecast population and ecosystem impacts, and form the basis of a bioeconomic model for Alaskan crab fisheries over a range of climate and ocean acidification scenarios (Sigler et al. 2008).

6.21 Red Sea

Relations between community calcification of an entire coral reef in the northern Red Sea and annual changes in temperature, aragonite saturation, and nutrient loading was investigated over a two-year period (Silverman et al. 2007). Calcification increased with temperature, with aragonite saturation state of reef seawater and with nutrients—which is in agreement with most laboratory studies and *in situ* measurements of single coral growth rates. Calcification rate also correlated well with precipitation rates of inorganic aragonite calculated for the same temperature and degree of saturation. Since only a minute portion of reef calcification is inorganic, these findings may be useful in predicting the response of coral reefs to ocean acidification and warming (Silverman et al. 2007).

6.22 Southern Ocean

Waters bordering Antarctica to 60° S latitude, also known as the Southern Ocean, are considered saturated with respect to CO_2 owing to recent climate changes (Le Quere et al. 2007). Between 1981 and 2004 the CO_2 sink has weakened by 0.08 petagrams of carbon per year per decade relative to the trend expected from the large increase in atmospheric CO_2 caused by an increase in Southern Ocean winds from human activities. Consequences include a reduction in the efficiency of the Southern Ocean sink of CO_2 in the short term (about 25 years), and a higher level of stabilization of atmospheric CO_2 on a multi-century time scale (Le Quere et al. 2007). However, Metzl et al. (2006) and Law et al. (2008) disagree with the findings of Le Quere et al. (2007). Metzl et al. (2006) aver that this region is neither a permanent CO_2 sink nor a permanent CO_2 source. Law et al. (2008) did not find a saturating Southern Ocean carbon sink due to recent climate changes. In their ocean model, wind forcing causes reduced carbon uptake, but heat and freshwater flux forcing cause increased uptake. Inversions of atmospheric CO_2 show that the Southern Ocean sink trend is dependent on network choice (Law et al. 2008).

Bottom waters of the Antarctic Ocean contain significantly less anthropogenic CO_2 than expected (Poisson and Chen 1987). Upwelled deep water from the Weddell Sea dilutes the anthropogenic CO_2 concentration in winter surface water which then mixes with the Weddell Shelf Water and more Weddell Deep Water to form the Antarctic Bottom water. The dilution of the winter surface water by the old Weddell Deep Water also explains why there is less excess CO_2 in winter water than in summer (Poisson and Chen 1987). Atmospheric inversion models to measure CO_2 uptake in the Southern hemisphere Ocean—based on initial oceanic CO_2 flux and atmospheric data—were unsuccessful owing, in part, to unreasonable land fluxes, inadequate representation of wintertime conditions, and mismatches between observed and simulated atmospheric CO_2. (Roy et al. 2003). If atmospheric CO_2

levels continue to increase at expected rates, the coldest surface waters in the Southern Ocean will become undersaturated with respect to aragonite (a metastable form of calcium carbonate) within 50 years (Orr et al. 2006). Southern Ocean acidification via anthropogenic CO_2 uptake is expected to adversely affect multiple calcifying species by lowering the concentration of carbonate ion to levels where aragonite and calcite shells begin to dissolve (McNeil and Matear 2008). Aragonite undersaturation is likely to occur when atmospheric CO_2 levels exceed 450 ppm, an event predicted to occur by the year 2030 and not later than 2038. Some prominent calcifying species, in particular veligers of the pteropod mollusc *Limacina helicina*, will experience detrimental carbonate conditions much earlier than previously thought (McNeil and Matear 2008).

The standing stock of water column particulate organic carbon (POC) and fluxes in the Ross Sea, Antarctica, in 1996–1997 is significantly affected by season of collection (Gardner et al. 2000). POC at 100 m during mid-October (early spring) was 240 mmol C/m^2, but POC more than doubled to 560 mmol C/m^2 10 days later. By mid-January (summer) the standing stock had increased to about 5,300 mmol C/m^2, but dropped to 3,500 mmol C/m^2 one week later. By late April (autumn) the standing stock was only 200 mmol C/m^2. The following spring the standing stock increased from 700 in late November to 2,200 mmol C/m^2 in early December. Variables affecting POC fluctuations include light, feeding populations of pteropods, resuspension and lateral advection (Gardner et al. 2000). Badly preserved aragonitic shells of the pteropod *Limacina helicina* were collected from deep (880 m) sediment traps moored in Terra Nova Bay (Manno et al. 2007). The saturation level for aragonite was located at about 1,000 m depth, or 200 m below the trap level. Aragonite dissolution above the chemical lysocline was attributed to: 1/low fluxes of of *Limacina helicina* near the bottom despite the large abundance of these pteropods in the upper 200 m; 2/ shells collected near the bottom showed a state of advanced chemical degradation, in contrast to shells collected by the shallow trap which appeared intact and well preserved; and 3/carbonate fluxes observed in the bottom trap corresponded to only 1% of fluxes measured in the shallow one. These results suggest an underestimation of the $CaCO_3$ budget in the deepest waters, with implications for the mechanisms influencing the inorganic carbon cycle (Manno et al. 2007).

Planktonic foraminiferans are single-celled calcite-secreting organisms that represent between 25 and 50% of the total open-ocean marine carbonate flux and influence the transport of organic carbon to the ocean interior (Moy et al. 2009). A comparison of shell weights of the modern foraminiferan *Globigerina bulloides* collected from sediment traps in the Southern Ocean with the weights of shells preserved in the underlying Holocene-aged sediments show that modern shell weights are 30% to 35% lower than those from the sediments, and this is consistent with reduced calcification today induced by ocean acidification. Authors also found a link between higher atmospheric carbon dioxide and low shell weights in a 50,000-year-long record obtained from a Southern Ocean marine sediment core. It is unclear whether reduced calcification will affect the survival of this and other species, but a decline in the abundance of foraminifera caused by acidification could affect marine ecosystems and the oceanic uptake of atmospheric carbon dioxide (Moy et al. 2009).

6.23 Tatoosh Island, Washington

A high resolution dataset spanning 8 years showed that pH declined with increasing atmospheric CO_2 levels but varied substantially in response to biological processes and physical conditions over multiple time scales (Wootton et al. 2008). For example, there is a pronounced 24-hour cycle spanning 0.24 pH units during a typical day This daily oscillation is explained by daily variation in photosynthesis and background respiration: water pH increases as CO_2 is taken up, by photosynthesis by day, and declines as respiration and diffusion from the atmosphere replenish CO_2 overnight. Also, pH fluctuated substantially among days and years, ranging up to 1.5 units over the study period. When the entire temporal span of these data were examined, a general declining trend in pH became apparent. The rate in decline of pH is substantially faster than that predicted by current models (Wootton et al. 2008). Nitrate and phosphorus estimates at Tatoosh Island were positively correlated with upwelling and negatively with sea surface temperature (Pfister et al. 2007). Ammonium and nitrite, however, were not correlated with upwelling or sea surface temperatures and showed elevated levels immediately adjacent to Tatoosh Island, suggesting major effects of local marine invertebrates, birds, and mammals on nutrient dynamics and cycling in coastal ecosystems (Pfister et al. 2007).

6.24 Literature Cited

Andersson, A.J., N.R. Bates, and F.T. Mackenzie. 2007. Dissolution of carbonate sediments under rising pCO_2 and ocean acidification: observations from Devil's Hole, Bermuda, *Aquat. Geochem.*, 13, 237–264.

Andruleit, H.A. 2000. Dissolution-affected coccolithophore fluxes in the central Greenland Sea (1994/1995), *Deep-Sea Res., II*, 47, 1719–1742.

Attrill, M.J., J. Wright, and E. Martin. 2007. Climate-related increases in jelly fish frequency suggest a more gelatinous future for the North Sea, *Limnol. Ocean*, 52, 480–485.

Balch, W.M., D.T. Drapeau, B.C. Bowler, E.S. Booth, L.A. Windecker, and A. Ashe. 2008. Space–time variability of carbon standing stocks and fixation rates in the Gulf of Maine, along the GNATS transect between Portland, ME, USA, and Yarmouth, Nova Scotia, Canada, *J. Plankton Res.*, 30, 119–139.

Balch, W.M., P.M. Holligan, and K.A. Kilpatrick. 1992. Calcification, photosynthesis and growth of the bloom-forming coccolithophore, *Emiliana huxleyi, Cont. Shelf Res.*, 12, 1353–1374.

Balch, W.M. and K. Kilpatrick. 1996. Calcification rates in the equatorial Pacific along 140 °W, *Deep-Sea Res., II*, 43, 971–993.

Bates, N.R. 2001. Interannual variability of oceanic CO_2 and biogeochemical properties in the Western North Atlantic subtropical gyre, *Deep-Sea Res., II*, 48, 1507–1528.

Bates, N.R. 2007. Interannual variability of the oceanic CO_2 sink in the subtropical gyre of the North Atlantic Ocean over the last 2 decades, *J. Geophys. Res.*, 112, C09013, doi:10.1029/2006JC003759.

Bates, N.R. and A.J. Peters. 2007. The contribution of atmospheric acid deposition to ocean acidification in the subtropical North Atlantic Ocean, *Mar. Chem.*, 107, 547–558..

Blackford, J.C. and F.J. Gilbert. 2007. pH variability and CO_2 induced acidification in the North Sea, *J. Mar. Systems*, 64, 229–242.

Borges, A.V. and M. Frankignoulle. 1999. Daily and seasonal variations of the partial pressure of CO_2 in surface seawater along Belgian and southern Dutch coastal areas, *J. Mar. System.*, 19, 251–266.

Brewer, P.G., C. Goyet, and G. Friederich. 1997. Direct observation of the oceanic CO_2 increase revisited, *Proc. Natl. Acad. Sci. USA*, 94, 8308–8313.

Broecker, W.S. and E. Clark. 2002. Carbonate ion concentration in glacial-age deep waters of the Caribbean Sea, *Geochem. Geophys. Geosyst.*, 3, 1021, doi:10.1029/2001GC000231.

Caldeira K., and M.E. Wickett. 2005 Ocean model predictions of chemistry changes from carbon dioxide emissions to the atmosphere and ocean, *J. Geophys. Res.*, 110, C09S04, doi:10.1029/2004JC002671.

Chisholm, J.R.M. 2000. Calcification by crustose coralline algae on the northern Great Barrier Reef, Australia, *Limnol. Ocean.*, 45, 1476–1484.

Cooper, D.J., A.J. Watson, and R.D. Ling. 1998. Variation of PCO_2 along a North Atlantic shipping route (U.K. to the Caribbean): a year of automated observations, *Mar. Chem.*, 60, 147–164.

Cooper, T.F., G. De'ath, K.A. Fabricius, and J.M. Lough. 2008. Declining coral calcification in massive *Porites* in two nearshore regions of the northern Great Barrier Reef, *Global Change Biol.*, 14, 529–538.

De'ath, G., J.M. Lough, and K.E. Fabricius. 2009. Declining coral calcification on the Great Barrier reef, *Science*, 323, 116–119.

Dore, J.E., R. Lukas, D.W. Sadler, M.J. Church, and D.M. Karl. 2009. Physical and biogeochemical modulation of ocean acidification in the central North Pacific, *Proc. Natl. Acad. Sci. USA*, 106, 12235–12240.

Feely, R.A., C.L. Sabine, J.M. Hernandez-Ayon, D. Ianson, and B. Hales. 2008. Evidence for upwelling of corrosive "acidified" water onto the continental shelf, *Science*, 320 (5882), 1490–1492.

Feng, Y., C.E. Hare, K. Leblanc, J.M. Rose, Y. Zhang, G.R. DiTullio, P.A. Lee, S.W. Wilhelm, J.M. Rowe, J. Sun, C. Gueguen, U. Passow, I. Benner, C. Brown, and D.A. Hutchins. 2009. Effects of increased pCO_2 and temperature on the North Atlantic spring bloom. I. The phytoplankton community and biogeochemical response, *Mar. Ecol. Prog. Ser.*, 388, 13–25.

Gardner, W.F., M.J. Richardson, and W.O. Smith Jr. 2000. Seasonal patterns of water column particulate organic carbon and fluxes in the Ross Sea, Antarctica, *Deep-Sea Res., II*, 47, 3423–3449.

Gledhill, D.K., R. Wanninkhof, F.J. Millero, and M. Eakin. 2008. Ocean acidification of the greater Caribbean region 1996–2006, *J. Geophys. Res.*, 113, C10031, doi:10.1029/2007JC004629.

Graziano, L.M., W.M. Balch, D. Drapeau. B.C, Bowler. R. Vaillancourt, and S. Dunford. 2000. Organic and inorganic carbon production in the Gulf of Maine, *Contin. Shelf Res.*, 20, 685–705.

Gruber, G. 1998. Anthropogenic CO_2 in the Atlantic Ocean, *Global Biogeochem. Cycles*, 12, 165–191, doi:10.1029/97GB03658.

Hall-Spencer, J.M., R. Rodolfo-Metalpa, S. Martin, E. Ransome, M. Fine, S.M. Turner, S.J. Rowley, D. Tedesco, and M.C. Buia. 2008. Volcanic carbon dioxide vents show ecosystem effects of ocean acidification, *Nature*, 454, 96–99.

Hare, C.E., K. Leblanc, G.R. DiTullio, R.M Kudela, Y. Zhang, P.A. Lee, S. Riseman, and D.A. Hutchins. 2007. Consequences of increased temperature and CO_2 for phytoplankton community structure in the Bering Sea, *Mar. Ecol. Prog. Ser.*, 352, 9–16.

Honjo, S., J. Dymond, R. Collier, and S.J. Manganini. 1995. Export production of particles to the interior of the equatorial Pacific Ocean during the 1992 EqPac experiment, *Deep-Sea Res. II*, 42, 831–870.

Joos, F., G.K. Plattner, T.F. Stocker, O. Marchal, and A. Schmittner. 1999. Global warming and marine carbon cycle feedbacks on future atmospheric CO_2, *Science*, 284, 464–467.

Jutterstrom, S. and L.G. Anderson. 2005. The saturation of calcite and aragonite in the Arctic Ocean, *Mar. Chem.*, 94, 101–110.

Kortzinger, A., U. Send, D.W.R. Wallace, J. Karstensen, and M. DeGrandpre. 2008. Seasonal cycle of O_2 and pCO_2 in the central Labrador Sea: atmospheric, biological, and physical implications, *Global Biogeochem. Cycles*, 22, GB1014. Doi:10.1029/2007GB003029.

Law, R.M., R.J. Matear, and R.J. Francey. 2008. Comment on "Saturation of the Southern Ocean CO_2 Sink due to Recent Climate Change", *Science*, 319, 570a.

Lee, K., S.D. Choi, G.H. Park, R. Wanninkhof, T.H. Peng, R.M. Key, C.L. Sabine, R.A. Feely, J.L. Bullister, F.J. Millero, and A. Kozyr. 2003. An updated anthropogenic CO_2 inventory in the Atlantic Ocean, *Global Biogeochem. Cycles*, 17(4), 1116, doi:10.1029/2003GB002067.

Lee, K., F.J. Millero, and R. Wanninkhof. 1997. The carbon dioxide system in the Atlantic Ocean, *J. Geophys. Res.*, 102(C7), 15,693–15707, doi:10.1029/97JC00067.

Le Quere, C., C. Rodenbeck, E.T. Buitenhuis, T.J. Conway, R. Langenfelds, A. Gomez, C. Labuschagne, M. Ramonet, T. Nakazawa, N. Metzi, N. Gillett, and M. Heimann. 2007. Saturation of the Southern Ocean CO_2 sink due to recent climate change, *Science*, 316, 1735–1738.

Letelier, RM., J.E. Dore, C.D. Winn, and D.M. Karl. 1996. Seasonal and interannual variations in photosynthetic carbon assimilation at Station ALOHA, *Deep-Sea Res., II*, 43, 467–490.

Manabe, S. and R.J. Stouffer. 1994. Multiple-century response of a coupled ocean-atmosphere model to an increase of atmospheric carbon dioxide, *J. Climate*, 7, 5–23.

Manno, C., S. Sandrini, L. Tositti, and A. Accornero. 2007. First stages of degradation of *Limacina helicina* shells observed above the aragonite chemical lysocline in Terra Nova Bay (Antarctica), *J. Mar. Syst.*, 68, 91–102.

Marshall, D.J., J.H. Santos, K.M.Y. Leung, and W.H. Chak. 2008. Correlations between gastropod shell dissolution and water chemical properties in a tropical estuary, *Mar. Environ. Res.*, 66, 422–429.

Martin, J.H., G.A. Knauer, D.M. Karl, and W.M. Broenkow. 1987. VERTEX: carbon cycling in the northeast Pacific, *Deep-Sea Res.*, 34, 267–285.

McNeil, B.I. and R.J. Matear. 2008. Southern Ocean acidification: a tipping pont at 450 ppm atmospheric CO_2, *Proc. Natl. Acad. USA*, 105, 18860–18864.

Metzl, N., C. Brunet, A. Jabaud-Jan, A. Poisson, and B. Schauer. 2006. Summer and winter air-sea CO_2 fluxes in the Southern Ocean, *Deep-Sea Res. I*, 53, 1548–2563.

Moy, A.D., W.R. Howard, S.G. Bray, and T.W. Trull. 2009. Reduced calcification in modern Southern Ocean planktonic foraminifera, *Nature Geosci.*, 2, 276–280.

Orr, J.C., L.G. Anderson, N.R. Bates, L. Bopp, V.J. Fabry, E. Jones, and D. Swingedouw. 2006. Arctic Ocean acidification, *EOS, Trans. Amer. Geophys. Union*, 87 (36), Suppl.

Orr, J.C., V.J. Fabry, O. Aumont, L. Bopp, S.C. Doney, R.A. Feely, A. Gnanadesikan, N. Gruber, A. Ishida, F. Joos, R.M. Key, K. Lindsay, E. Maier-Reimer, R. Matear, P. Monfray, A. Mouchet, R.G. Najjar, G.K. Plattner, K.R. Rodgers, C.L. Sabine, J.L. Sarmiento, R. Schlitzer, R.D. Slater, I.J. Totterdell, M.F. Weirig, Y. Yamanaka, and A. Yool. 2005. Anthropogenic ocean acidification over the twenty-first century and its impact on calcifying organisms, *Nature*, 437 (7059), 681–686.

Pfister, C.A., J.T. Wooton, and C.J. Neufeld. 2007. The relative roles of coastal and oceanic processes in determining physical and chemical characteristics of an intensively sampled nearshore system, *Limnol. Ocean*, 52, 1767–1775.

Ploug, H. 2008. Cyanobacterial surface blooms formed by *Aphanizomenon* sp. and *Nodularia spumigena* in the Baltic Sea: small-scale fluxes, pH, and oxygen microenvironments, *Limnol. Ocean*, 53, 914–921.

Poisson, A. and C.A. Chen. 1987. Why is there little anthropogenic CO_2 in the Antarctic bottom water?, *Deep Sea Res. Part A*, 34, 1255–1275.

Poulton, A.J., P.M. Holligan, A. Hickman, Y.N. Kim, T.R. Adey, M.C. Stinchcombe, C. Holeton, S. Root, and E.M..S. Woodward. 2006a. Phytoplankton carbon fixation, chlorophyll-biomass and diagnostic pigments in the Atlantic Ocean, *Deep-Sea Res. II*, 53, 1593–1610.

Poulton, A.J., R. Sanders, P.M. Holligan, M.C. Stinchcombe, T.R. Adey, L. Brown, and K. Chamberlain. 2006. Phytoplankton mineralization in the tropical and subtropical Atlantic Ocean, *Global Biogeochem. Cycles*, 20, GB4002, doi:10.1029/2006GB002712.

Rodolfo-Metalpa, R., C. Lombardi, S. Cocito, J.M. Hall-Spencer, and M.C. Gambi. 2010. Effects of ocean acidification and high temperatures on the bryozoan *Myriapora truncata* at natural CO_2 vents, *Mar. Ecol.*, 31, 447–456.

Roy, T., P. Rayner, R. Matear, and R. Francey. 2003. Southern hemisphere ocean CO_2 uptake: reconciling atmospheric and oceanic estimates, *Tellus*, 55B, 701–710.

Ruhl, H.A., J.A. Ellena, and K.L. Smith Jr. 2008. Connections between climate, food limitation, and carbon cycling in abyssal sediment communities, *Proc. Natl. Acad. Sci. USA*, 195, 17006–17011.

Sabine, C.L., R.A. Feely, R.M. Key, J.L. Bullister, F.J. Millero, K. Lee, T.H. Peng, B. Tilbrook, T. Ono, and C.S. Wong. 2002a. Distribution of anthropogenic CO_2 in the Pacific Ocean, *Global Biogeochem. Cycles*, 16(4), 1083, doi:10.1029/2001GB001639.

Sabine, C.L., R.M. Key, R.A. Feely, and D. Greeley. 2002. Inorganic carbon in the Indian Ocean: distribution and dissolution processes, *Global Biogeochem. Cycles*, 16(4). 1067, doi:10.1029/2002GB001869.

Sabine, C.L., R.M. Key, K.M. Johnson, J. Millero, A. Poisson, J.L. Sarmiento, D.W.R. Wallace, and C.D. Winn. 1999. Anthropogenic CO_2 inventory of the Indian Ocean, *Global Biogeochem. Cycles*, 13, 179–198, doi:10.1029/1998GB900022.

Schuster, U. and A.J. Watson. 2007. A variable and decreasing sink for atmospheric CO_2 in the North Atlantic, *J. Geophys. Res.*, 112, C11006, doi:10.1029/2006JC003941.

Sigler, M.F., R.J. Foy, J.W. Short, M. Dalton, L.B. Eisner, T.P. Hurst, J.F. Morado, and R.P. Stone. 2008. Forecast fish, shellfish and coral population responses to ocean acidification in the north Pacific Ocean and Bering Sea: an ocean acidification research plan for the Alaska Fisheries Science Center, *AFSC process. Rep. 2008–07*, 35 pp. Avail. from NOAA, NMFS, 17109 Point Lena Loop Road, Juneau, AK 99801.

Silverman, J., B. Lazar, and J. Erez. 2007. Effect of aragonite saturation, temperature, and nutrients on the community calcification rate of a coral reef, *J. Geophys. Res.*, 112, C05004, doi:10.1029/2006JC003770.

Smith, K.L. Jr., R.J. Baldwin, H.A. Ruhl, M. Kahru, B.G. Mitchell, and R.S. Kaufmann. 2006. Climate effect on food supply to depths greater than 4,000 meters in the northeast Pacific, *Limnol. Ocean*, 51, 166–176.

Smith, K.L. Jr., R.S. Kaufmann, R.J. Baldwin, and A.F. Carlucci. 2001. Pelagic-benthic coupling in the abyssal eastern North Pacific: an 8-year time series study of food supply and demand, *Limnol. Ocean*, 46, 543–556.

Somasundar, K., A. Rajendran, M.D. Kumar, and R.S. Gupta. 1990. Carbon and nitrogen budgets of the Arabian Sea, *Mar. Chem.*, 30, 363–377.

Steinacher, M., F. Joos, T.L. Frolicher, G.K. Plattner, and S.C. Doney. 2009. Imminent ocean acidification in the Arctic projected with the NCAR global couple carbon cycle-climate model, *Biogeosciences*, 6, 515–533.

Takahashi, K., N. Fujitani, M. Yanada, and Y. Maita. 2000. Long-term biogenic particle fluxes in the Bering Sea and the central subarctic Pacific Ocean, 1990–1995, *Deep-Sea Res. I* 47, 1723–1759.

Takahashi, T., S.C. Sutherland, R.A. Felly, and R. Wanninkhof. 2006. Decadal change of the surface water pCO_2 in the North Pacific: a synthesis of 35 years of observations, *J. Geophys. Res.*, 111, C07805, doi:10.1029/2005JC003074,2006.

Takahashi, T., S.C. Sutherland, C. Sweeney, A. Poisson, N. Metzl, B. Tilbrook, N. Bates, R. Wanninkhof, R.A. Feely, C. Sabine, J. Olafsson, and Y. Nojiri. 2002. Global sea-air CO_2 flux based on climatological surface ocean pCO_2 and seasonal biological and temperature effects, *Deep-Sea Res., II* 49, 1601–1622.

Thomas, H., Y. Bozec, H.J.W. de Baar, K. Elkalay, M. Frankignoulle, L.S. Schiettecatte, G. Kattner, and A.V. Borges. 2005. The carbon budget of the North Sea, *Biogeosciences*, 2, 87–96.

Tortell, P.D., G.R. DiTullio, D.M. Sigman, and F.M.M. Morel. 2002. CO_2 effects on taxonomic composition and nutrient utilization in an Equatorial Pacific phytoplankton assemblage, *Mar. Ecol. Prog. Ser.*, 236, 37–43.

Tsunogai, S. and S. Noriki. 1991. Particulate fluxes of carbonate and organic carbon in the ocean. Is the marine biological activity working as a sink of the atmospheric carbon? *Tellus*, 43B, 256–266.

Tunnicliffe, V., K.T.A. Davies, D.A. Butterfield, R.W. Embley, J.M. Rose, and W.W. Chadwick Jr. 2009. Survival of mussels in extremely acidic waters on a submarine volcano, *Nature Geosci.*, 2, 344–348.

Watson, A.J., U. Schuster, D.C.E. Baker, N.R. Bates, A. Corbiere, M. Gonzalez-Davila, T. Friedrich, J. Hauck, V. Heinze, T. Johannessen, A. Kortsinger, N. Metzl, J. Olafsson, and others. 2009. Tracking the variable North Atlantic sink for atmospheric CO_2, *Science*, 326, 1391–1393.

Wootton, J.T., C.A. Pfister, and J.D. Forester. 2008. Dynamic patterns and ecological impacts of declining ocean pH in a high-resolution multi-year dataset, *Proc. Natl Acad. Sci. USA*, 105, 18848–18853.

Yahel, G., F. Whitney, H.N. Reiswig, D.I. Eerkes-Medrano, and S.P. Leys. 2007. *In situ* feeding and metabolism of glass sponges (Hexactinellida, Porifera) studied in a deep temperate fjord with a remotely operated submersible, *Limnol. Ocean*, 52, 428–440.

Zhang, X. and H.G. Dam. 1997. Downward export of carbon by diel migrant mesozooplankton in the central equatorial Pacific, *Deep-Sea Res. II*, 44, 2191–2202.

CHAPTER 7

Modifiers

7.1 General

Natural variations in seawater pH and atmospheric carbon dioxide levels are documented that preclude anthropogenic input. Moreover, some methods for measuring oceanic carbon are suspect. Overall, there is a huge uncertainty as to what extent organism adaptation or acclimatization will mitigate the long term effects of ocean acidification (Widdicombe and Spicer 2008). Interactions with other elements, latitudes, temperature, and depth are also documented (Andersson et al. 2008). For example, factors which determine the mineralogy, composition, and texture and rate of crystallization of carbonate precipitates from seawater or marine pore waters are significantly modified by the Mg^{2+}/Ca^{2+} ratio of the solution, presence of orthophosphate ions, sulfate concentration of the fluids, water temperature, and salinity (Zhong and Mucci 1989). These modifiers—and others—must be considered in any review of human-caused oceanic acidification from combustion of fossil fuels.

7.2 Methodological

Rothman (2002) concluded that degassing and silicate weathering were the primary controls on the carbon cycle for the last 500 million years and that $p\text{CO}_2$ does not exert dominant control of Earth's climate at time scales greater than 10 million years. But the results do not indicate whether either of these mechanisms dominated, or if weathering was driven by the diversification of land plants, continental collisions, or a complex combination of tectonic, biological, and geochemical processes (Rothman 2002).

The Mg/Ca ratio in tests of modern echinoids is about 5.0, whereas this value was <2.0 in tests of fossil echinoderms 550 million years ago (Dickson 2002). Seawater chemistry over the past 550 million years, as reflected by changes in marine nonskeletal carbonate mineralogy, specifically the carbon molar ratio in carbonate mineral precipitation, has been advanced by Montanez (2002). The hypothesis that seawater Mg/Ca ratio has been the primary influence on the mineralogy of marine chemical precipitates is, however, controversial given that it requires substantial change in the Mg/Ca ratio of paleo-oceans, for which there is no direct evidence. Also there could be significant uncertainty associated with using the skeletal mineralogy and composition of fossil organisms as an environmental indicator of seawater Mg/Ca ratios. Author concludes that the influence of fluctuation in seawater Mg/Ca ratios on the carbonate productivity of major sediment producers, such as calcareous nannoplankton for deep-water chalks, and the biologic composition of shallow water reef builders has major implications for global carbon cycling and for atmospheric CO_2 levels (Montanez 2002). Changes in seawater Mg/Ca remain one of the largest uncertainties in using Mg paleothermometry to determine bottom water temperatures over long (tens of millions of years) time scales (Lear et al. 2002). Comparison of calibration data with benthic foraminiferal Mg/Ca and ^{18}O indicates that seawater Mg/Ca was not more than 35% lower 49 million years ago than today (Lear et al. 2002). But seawater Mg/Ca ratios have varied significantly throughout the eons from a low of 1.0 to the present 5.0 (Ries 2006). Crustose coralline alga, *Neogoniolithon* sp., held in seawater treatments formulated with identical Mg/Ca ratios—but differing absolute concentrations of Mg and Ca—showed no differences in skeletal Mg/Ca ratios, suggesting that the Mg/Ca ratio is more important than the absolute concentration of Mg in determining their skeletal composition. Empirical fossil evidence indicates that variations of oceanic Mg/Ca ratios has caused the mineralogy and skeletal chemistry of many calcifying marine organisms to change significantly over geologic time (Ries 2006).

The net exchange of carbon dioxide between the atmosphere and the ocean is dominated by seasonal dynamics of carbon cycling in the upper ocean (Michaels et al. 1994). This cycle is a balance between abiotic and biotic carbon transport from the ocean's upper layer. Measurement of these processes were made over five years in the Sargasso Sea off Bermuda (Michaels et al. 1994). The decrease in carbon stocks from spring to autumn in the upper 150 m of the ocean is three times larger than the measured sum of biotic and biotic fluxes out of this layer. This discrepancy is attributed to failure to account for horizontal advection or to inaccuracies in the fluxes of sinking particles as measured by sediment traps. Either the traps miss 80% of the sinking particles, or 70% of the carbon cycling is due to advection. Sediment-trap measurements of the flux during this period suggest that most of the discrepancy may be due to inaccuracies in the trap methods, which would require a reassessment of particle export and remineralization of carbon in the oceans. If advection is the main source of the discrepancy, the traditional vertical modeling of the oceanic carbon cycle cannot give a full account of carbon dynamics (Michaels et al. 1994). Strong seasonal patterns in total carbon dioxide and alkalinity were observed in the Sargasso Sea in February 1992 and could be due to open-ocean calcification by carbonate-secreting organisms rather than to physical processes (Bates et al. 1996). Coccolithophore calcification is the most likely cause of this event, although calcification by foraminiferans or pteropods can not be precluded. Authors aver that open-ocean calcification and biological community structure are both important in the biogeochemical cycling of carbon (Bates et al. 1996).

7.3 Natural Variations

Systematic fluctuations in seawater chemistry over the past 540 million years are in phase with oscillations in sea floor spreading rates, volcanism, global sea level, and the primary mineralogies of marine limestones and evaporites (Lowenstein et al. 2001). During the Late Precambrian (about 540 million years ago), Permian (255 million years ago), and Tertiary to the present (0–40 million years ago), seawater had high Mg/Ca ratios (>2.5), and relatively high sodium concentrations when aragonite and $MgSO_4$ salts were the dominant marine precipitates. Conversely, seawater had low Mg/Ca ratios (<2.3) and relatively low sodium concentrations during the Cambrian (540–520 million year ago), Silurian (440–418 MYA) and Cretaceous (124–94 MYA) when calcite was the dominant nonskeletal carbonate and K-Mg- and Ca-bearing chloride salts were the only potash evaporites (Lowenstein et al. 2001). Based on oxygen isotopes in calcite and aragonite shells, tropical sea surface temperatures over the past 550 million years show large oscillations in sea surface temperatures with the cold-warm cycles (Velzer et al. 2000). These data conflict with a temperature reconstruction using an energy balance based on reconstructed carbon dioxide concentrations. The results, however, can be reconciled if atmospheric carbon dioxide concentrations were not the principal driver of climate variability for at least one third of this time period, or if the reconstructed carbon dioxide concentrations are not reliable (Velzer et al. 2000).

A 300-million year record of atmospheric carbon dioxide concentrations from fossil plant cuticles indicates that CO_2 levels for much of this period was greater than 1,000 ppm with only two intervals of low CO_2 (<1,000 ppm) which coincide with known ice ages (Retallack 2001). Atmospheric carbon dioxide concentrations in the early Cenozoic era—about 60 million years ago—of more than 2,000 ppm were estimated based on boron-isotope ratios of ancient foraminiferan shells (Pearson and Palmer 2000). The boron-isotope approach to CO_2 estimation relies on the fact that a rise in the atmospheric concentration will mean that more CO_2 is dissolved in the surface ocean, causing a reduction in pH. The pH of ancient sea water was estimated by measuring the boron-isotope composition of calcium carbonate precipitated from it. This is because boron in aqueous solutions occurs as two chemical species, $B(OH)_3$ and $B(OH)_4^-$, between which the equilibrium is strongly pH-dependent over the natural acidity range of sea water (as quoted in Pearson and Palmer 2000).

About 55 million years ago, The Paleocene/Eocene thermal maximum was a brief period of widespread extreme climatic warming that was associated with massive atmospheric greenhouse gas input (Sluijs et al. 2006). Maximum sea surface temperatures near the North Pole increased from about 18°C to over 23°C during this event, implying the absence of ice and a rise in sea level. Stable carbon isotope records of carbonate and organic carbon show about a 2.5% drop, suggesting an input of at least 1.5×10^{18} g of ^{13}C-depleted carbon or somewhat analogous in magnitude and composition to current and expected fuel emissions (Sluijs et al. 2006). Intrusion of mantle-derived melts in carbon-rich sedimentary strata in the northeast Atlantic may have caused an explosive release of methane, transported to the ocean or atmosphere through the vent complexes, close to the Paleocene Eocene boundary (Svensen et al. 2004). Similar volcanic and metamorphic processes may explain climate events associated with other large igneous provinces of 183–250 million years ago (Svensen et al. 2004). The rapid release of 200,000 million tons of carbon in the form of methane (source unknown) during the Paleocene-Eocene period lowered deep-sea pH, triggering a rapid (<10,000 years) shoaling of the calcite compensation depth followed by gradual recovery of >100,000 years (Zachos et al. 2005). Regardless of the source, the released methane was rapidly oxidized to CO_2. Subsequent dissolution of this CO_2 would lower the pH and carbonate ion content of seawater. Field studies in the South Atlantic Ocean indicated that a large mass of carbon dissolved at the Paleocene-Eocene boundary was permanently sequestered through silicate weathering feedback. The lysocline, also referred to as the calcite saturation horizon, represents the depth in the ocean where the carbonate ion concentration falls below the saturation level, currently about 4 km in the South Atlantic Ocean. Carbonate accumulation can occur below this level if the flux of carbonate to the sea floor exceeds the rate at which it dissolves. The depth at which dissolution is greater than the flux and where carbonate does not accumulate defines the calcite compensation depth (Zachos et al. 2005).

Between 40 and 55 million years ago, CO_2 levels erratically declined to 700–900 ppm, owing, perhaps to reduced out gassing from ocean ridges, volcanoes, and increased carbon burial. About 24 million years ago, atmospheric CO_2 levels remained below 500 ppm and were more stable than before, although transient intervals of CO_2 reduction may have occurred during periods of rapid cooling approximately 3 to 25 million years ago. A global cooling trend about 2 to 4 million years ago caused a reduction in atmospheric CO_2 levels from 280 to 210 ppm (Pearson and Palmer 2000). During the Miocene Climactic Optimum of about 14.5–17 million years ago, fossil flora and faunal evidence indicated that this was the warmest of the past 35 million years with mid-latitude temperatures as much as 6°C higher than at present (Flower 1999). However, low CO_2 levels of about 180–290 ppm throughout the early to late Miocene (9–25 million years ago) indicate that high

CO_2 cannot explain Miocene warmth, and that other mechanisms must have had a greater influence (as quoted in Flower 1999).

Ice cores from Antarctica provide a composite record of atmospheric carbon dioxide over the past 650,000 years (Luthi et al. 2008). Analysis of cores to a depth of 200 m extends the record of two complete glacial cycles to 800,000 years before the present. Antarctic atmospheric carbon dioxide is strongly correlated with Antarctic temperature throughout eight glacial cycles, but with significantly lower concentrations of 172 to 300 ppm CO_2 and below 180 ppm for a period of 3,000 years during the late Quaternary (Luthi et al. 2008). Wind-driven upwelling in the ocean around Antarctica helps regulate the exchange of CO_2 between the deep sea and the atmosphere, as well the supply of dissolved silicon to the euphotic zone of the Southern Ocean (Anderson et al. 2009). Diatom productivity south of the Antarctic Polar Front and the subsequent burial of biogenic opal in underlying sediments are limited by this silicon supply. Opal burial rates, and thus upwelling, were enhanced during the termination of the last ice age in each sector of the Southern Ocean. Evidence is presented for two intervals of enhanced upwelling concurrent with the two intervals of rising atmospheric CO_2 during deglaciation. Authors conclude that these results directly link increased ventilation of deep water to the deglacial rise in atmospheric CO_2 (Anderson et al. 2009).

A history of glacial to Holocene radiocarbon in the deep western North Atlantic Ocean from deep sea corals and benthic foraminifera shows that deglaciation is marked by switches between radiocarbon-enriched and radiocarbon-depleted waters (Robinson et al. 2005). The large radiocarbon gradients significantly modify atmospheric radiocarbon. The deep ocean radiocarbon pattern supports the notion of the bipolar seesaw. When the deep ocean was flushed by radiocarbon-rich northern source water, Greenland was warming. And when northern source water was replaced by southern source water, Greenland was cooling. The intermediate depth ocean is more variable with multiple switches between radiocarbon-depleted and radiocarbon-enriched water masses, but it is deep ocean variability that has the major role in modulating atmospheric carbon (Robinson et al. 2005). Boron-isotope composition of a 300-year-old massive *Porites* sp. coral in the southwestern Pacific Ocean was studied (Pelejero et al. 2005). Large variations in pH—almost 0.3 units—were found over 50-year cycles that varied with periodic oscillations of ocean-atmosphere anomalies. Authors suggest that natural pH cycles can modify the impact of ocean acidification on coral reef ecosystems (Pelejero et al. 2005).

7.4 Interactions

Few animals may be acutely sensitive to moderate CO_2 increases, but subtle changes may already have started to be felt in a wide range of species (Portner et al. 2005). Carbon dioxide effects identified in marine invertebrates from habitats characterized by oscillating CO_2 levels include depressed metabolic rates and reduced ion exchange and protein synthesis rates. During future climate change, simultaneous shifts in temperature, CO_2, and hypoxia levels will enhance sensitivity to environmental extremes relative to change in a single variable. These interactions need to be considered in terms of future increases in atmospheric CO_2 and its uptake by the ocean as well as in terms of proposed mitigation efforts. It is not now known to what extent or how quickly species may adapt to permanently elevated CO_2 levels by microevolutionary compensatory processes (Portner et al. 2005).

Ocean warming is expected to cause major shifts in the flow of carbon and energy through the pelagic system (Wohlers et al. 2009). Increasing temperature by 2 to 6°C decreased the biological drawdown of dissolved inorganic carbon by up to 31%. The loss of organic carbon through sinking was significantly reduced at elevated temperatures. The observed changes in biogenic carbon flow may inhibit the transfer of primary produced organic matter to higher trophic levels, weaken the ocean's biological carbon pump, and provide a positive feedback to rising atmospheric CO_2 (Wohlers et al. 2009). In a model of ocean-atmosphere interaction that excluded biological processes, the oceanic uptake of atmospheric CO_2 was substantially reduced with global warming owing to the collapse of the ocean thermohaline circulation, thereby directly impacting future growth rate of atmospheric CO_2 (Sarmiento and Le Quere 1996). Model simulations that include biological processes show a large offsetting effect from the downward flux of biogenic carbon, but authors were unable to quantify the magnitude of this effect owing to insufficient present knowledge (Sarmiento and Le Quere 1996). Ocean acidification is occurring alongside other climate-related stressors, such as ocean warming and sea-level rise (UNEP 2010). These are compounded by non-climate related impacts such as overfishing and pollution, which add pressure to already strained marine ecosystems that provide food for human consumption (UNEP 2010). Numerous factors can significantly modify projected effects of CO_2-induced oceanic acidification including other elements, depth, temperature, latitude, ecosystem composition, and chemical composition of seawater and sediments. For example, ecosystem responses to climate change include altered CO_2 patterns, hypoxia, salinity change, and eutrophication (Portner and Farrell 2008). Key to setting sensitivity to ocean acidification are the mechanisms and efficiency of systemic acid-base regulation. Specific effects of each stressor will reduce whole-organism performance, especially at extreme temperatures, thereby narrowing thermal windows and reducing biogeographical ranges. Studies of ecosystem consequences of stressors like ocean acidification through carbon dioxide should thus consider effects on thermally limited oxygen supply (Portner and Farrell 2008).

Interactive effects were measured of near-future ocean warming and acidification regimes on fertilization in intertidal and shallow subtidal echinoids (*Heliocidaris erythrogramma, Heliocidaris tuberculata, Tripneustes gratilla, Centrostephanus rodgersii*), an asteroid (*Patirella regularis*), and an abalone (*Haliotis coccoradiata*) found in Australia (Byrne et al. 2010). Temperatures as high as 26°C coupled with pH as low as 7.6 had no significant effect on fertilization in all six species. Results suggest that these species are either particularly resistant or able to adapt to changing seawater chemistry and temperature (Byrne et al. 2010). Interactive effects of warming and acidification on fertilization and development of *Heliocidaris erythrogramma*, an echinoid echinoderm, were measured in all combinations of 20° to 26°C and pH 7.6 to 8.2, with 20°C and pH 8.2 being ambient (Byrne et al. 2009). Percent fertilization was >89% across all treatments. Normal development occurred at all pH levels; however, at 24°C, cleavage was reduced by 40% and at 26°C by a further 20%. Normal gastrulation fell below 4% at 26°C, and development was impaired. Authors confirm the thermotolerance and pH resilience of fertilization and embryogenesis at predicted climate change scenarios, with negative effects at upper limits of ocean warming. Ocean acidification may impair calcification but embryos may not reach the skeletogenic stage in a warm ocean, suggesting that temperature and not pH is the more important stressor (Byrne et al. 2009). However, in a later paper (Byrne et al. 2010a), they indicate that decreased pH and increased CO_2 narcotizes sea urchin sperm indicating that acidification may indeed impair fertilization. In contrast, increased temperature may enhance fertilization. In a series of studies they found that: fertilization was positively affected by sperm density; increased acidification and CO_2 did not reduce fertilization, even at low sperm density; and that increased temperature did not enhance fertilization. They conclude that sea urchin fertilization is robust to climate change stressors, but developmental stages may be vulnerable to ocean change (Byrne et al. 2010a). Results of interactive effects of ocean warming and CO_2-driven acidification on development and calcification in the tropical sea urchin *Tripneustes gratilla* shows the positive and negative effects of these stressors (Brennand et al. 2010). Calcification was impaired because of decreased calcium carbonate saturation with decreased pH and a lowered metabolism owing to hypercapnia. However, these effects may be countered by enhanced growth and metabolism due to warming (Brennand et al. 2010). Ocean warming, acidification, decreased carbonate saturation, and their interactive effects are likely to impair skeletogenesis, with negative implications for many marine species (Byrne et al. 2010b). Interactive effects of warming and acidification on an abalone (*Haliotis coccoradiata*) and an echinoderm (*Heliocidaris erythrogramma*) reared from fertilization were observed under a variety of combined thermal and pH/CO_2 regimens. Exposure of these species to warming (+2° to +4°C) and acidification (pH 7.6–7.8) resulted in unshelled larvae and abnormal juveniles. Abalone development was most sensitive with no interaction between stressors. For sea urchin, the percent of abnormal larvae increased in response to both stressors, although a +2°C.warming diminished the effect of low pH. Authors conclude that projected near-future climate change will have deleterious effects on development with differences in vulnerability in the two species (Byrne et al. 2010b). Exposure of adults of the ophiuroid brittlestar, *Ophiura ophiura* for 40 days to reduced pH (7.3) and increased temperature (+4.5°C) predicted to occur by the year 2300 resulted in a 30% reduction in the rate of excised arm regeneration, suggesting that fitness and survival are reduced through slower recovery from arm damage (Wood et al. 2010). In the case of the jumbo squid, *Dosidecus gigas*, a top predator in the eastern Pacific

Ocean, a decrease in pH of 0.3 units will depress metabolic rates by 31% and activity levels by 45% (Rosa and Seibel 2008). This effect is exacerbated by high temperatures. Reduced aerobic and locomotory scope in warm, high-CO_2 surface waters will presumably impair predator-prey interactions with cascading consequences for growth, survival and reproduction. As the oxygen minimum layer shoals, squids will retreat to shallower, less hospitable, waters at night to feed and repay any oxygen debt accumulated during their vertical migration into the oxygen minimum layer. In the absence of adaptation or horizontal migration, the synergism between ocean acidification, global warming, and expanding hypoxia will compress the habitable depth range of this commercially and ecologically important predator (Rosa and Seibel 2008).

Seasonal variation of CO_2 and nutrients in the high-latitude surface ocean are documented (Takahashi et al. 1993). Spring phytoplankton blooms in the surface water of the North Atlantic Ocean and Iceland Sea, as one example, caused a marked reduction of surface water pCO_2 and the concentrations of CO_2 and nutrients within two weeks, and proceeded until the nutrient salts were exhausted. This type of seasonal behavior is limited to the high-latitude North Atlantic Ocean and adjoining seas. In contrast, seasonal changes in CO_2 and nutrients were more gradual in the North Pacific Ocean and the nutrients were only partially consumed in the surface waters of the subarctic North Pacific Ocean and Southern Ocean. In the subpolar and polar waters of the North and South Atlantic Ocean and the North Pacific Ocean, pCO_2 and the concentrations of CO_2 were much higher during winter than summer. During winter, the high latitude areas of the North Atlantic Ocean, North Pacific Ocean, and Weddell Sea were sources for atmospheric CO_2; during summer they became CO_2 sinks owing to the upwelling of deep waters rich in CO_2 and nutrients during winter and the intense photosynthesis occurring in strongly stratified upper layers during summer. However, subtropical waters were a CO_2 source in summer and a sink in winter owing to low nutrient levels and increasing summer temperatures. An intense CO_2 sink zone was found along the confluence of the subtropical and polar waters; its formation is attributed to the combined effects of cooling in subtropical waters and photosynthetic drawdown of CO_2 in subpolar waters (Takahashi et al. 1993).

Diurnal hourly changes in seawater from intertidal rockpools were considerable (Truchot and Duhamel-Jouve 1980). During night emersion periods, pH and oxygen content were low, alkalinity was high, and CO_2 increased. During day emersion periods, however, oxygen content increased, pH increased to more than 10, and CO_2 and alkalinity decreased. It appears that respiratory and photosynthetic activities were the main activities involved (Truchot and Duhamel-Jouve 1980). Changes in pH will change the organic and inorganic speciation of metals in surface ocean water, with increases or decreases of at least 50% expected between the year 2000 and the year 2100 (Table 7.1). Metals expected to increase include some salts of aluminum, gallium, beryllium , copper, uranium, lanthanum, cerium, praseodymium, promethium, neodymium, samarium, europium, gadolinium, terbium, dysprosium, holmium, erbium, thulium, ytterbium. lutetium, lead and yttrium (Table 7.1). Metals expect to decrease include hydroxide and carbonate salts of indium, beryllium, copper, lanthanum, cerium, praseodymium, neodymium, promethium, samarium, europium, iron, nickel, cobalt, and zinc (Table 7.1). The effect of these changes on the interactions of metals with marine organisms are examined, as is the decrease in concentration of OH^- and CO_3^{2-} ions on the solubility, adsorption, toxicity and rates of redox processes of metals in seawater (Millero et al. 2009).

The ocean carbon cycle is correlated with cycles of nitrogen, phosphorus, and silicon—with consequences for the microbes that mediate many key nutrient transformations (Hutchins et al. 2009). For example, the nitrogen cycle responds to higher CO_2 levels through increases in global N_2 fixation, denitrification, and potential increases in nitrification. The consequences could include reduced supplies of oxidized substrates to denitrifiers, lower levels of nitrate-supported new primary production, and shifts in current phytoplankton communities. The phosphorus and silicon cycles seem less likely to be directly enhanced by enhanced CO_2 conditions, but will respond indirectly to changing carbon and nitrogen biogeochemistry. Because collected results of experiments that manipulated CO_2 in complex natural plankton assemblages differ widely from those of monospecific culture studies, authors state that much more information is needed to confidently extrapolate incubation results to century-scale ocean-wide trends (Hutchins et al. 2009). Nitrogen system transformations and their effects on pH are common to all

Table 7.1. Metal salts[a] in surface seawater projected to undergo at least a 50% change between the year 2000 (pH 8.1) and 2100 (pH 7.7). Modified from Millero et al. (2009).

Metal species showing at least a 50% increase	Metal species showing at least a 50% decrease
$Al(OH)_3$	$In(OH)_4^-$
$Ga (OH)_3$ Be^{2+}	$Be(OH)_2$; $Be(OH)_3$
Cu^{2+}	$Cu(CO_3)_2^{2-}$
$UO_2(CO_3)_2^{2-}$	$La (CO_3)_2^-$
La^{3+}; $LaSO_4^-$; $LaCl^{2+}$	$Ce(CO_3)_2^-$
Ce^{3+}; $CeSO_4^+$; $CeCl^{2+}$	$Pr(CO_3)_2^-$
Pr^{3+}; $PrSO_4^+$; $PrCl_2^+$	$Nd(CO_3)_2^-$
Nd^{3+}; $NdSO_4^+$; $NdCl^{2+}$	$Pm(CO_3)_2^-$
Pm^{3+}; $PmSO_4^+$; $PmCl^{2+}$	$Sm(CO_3)_2^-$
Sm^{3+}; $SmSO_4^+$; $SmCl^{2+}$	$Eu(CO_3)_2^-$
Eu^{3+}; $EuSO_4^+$; $EuCl^{2+}$	$FeCO_3$; $FeOH$
Gd^{3+}; $GdSO_4^+$; $GdCl^{2+}$	$NiCO_3$
Tb^{3+}; $TbSO_4^+$; $TbCl^{2+}$	$CoCO_3$; $CoOH$
Dy^{3+}; $DySO_4^+$; $DyCl^{2+}$	$ZnOH^+$
Ho^{3+}; $HoCO_3^+$; $HoSO_4^+$	$ZnCO_3$
Er^{3+}; $ErCO_3^+$	
Tm^{3+}; $TmCO_3^+$	
Yb^{3+}; $YbCO_3^+$	
Lu^{3+}; $LuCO_3^+$	
Pb^{2+}; $PbCl^+$; $PbCl_2$; $PbCl_3^-$	
Y^{3+}; YCl^{2+}; YF^{2+}	

[a]Aluminum = Al; Beryllium = Be; Cerium = Ce; Copper = Cu; Dysprosium = Dy; Erbium = Er; Europium = Eu; Gadolinium = Gd; Holmium = Ho; Indium= In; Lanthanum = La; Lead = Pb; Lutetium = Lu; Neodymium = Nd; Nickel = Ni; Praseodymium = Pr; Promethium = Pm; Samarium = Sm; Terbium = Tb; Thulium = Tm; Uranium = U; Ytterbium = Yb; Yttrium = Y; Zinc = Zn.

aquatic systems and are demonstrated in three species of marine phytoplankton: *Phaeodactylum tricornutum, Dunaliella tertiolecta,* and *Monochrysis lutheri* in continuous culture (Brewer and Goldman 1976). Uptake of NO_3^- caused an increase in alkalinity, whereas uptake of NH_4^+ produced a decrease. Photosynthetic assimilation of inorganic nitrogen—in which NO_3^- uptake is balanced by OH^- production and NH_4^+ uptake leads to H^+ generation—suggests active uptake of nitrogen species by these organisms (Brewer and Goldman 1976). Blooms of the diatom *Skeletonema costatum* were initiated with P-deficient medium under two different initial concentrations of dissolved CO_2 (Gervais and Riebesell 2001). With growth decreasing due to phosphorus limitation, carbon isotope fractionation increased in both CO_2 treatments by 2–3%, despite decreasing CO_2. Systematic changes in algal growth during blooms affect isotope fractionation, severely complicating the interpretation of carbon isotope measurements in suspended and sedimentary organic matter (Gervais and Riebesell 2001).

Some species of coccolithophorids precipitate calcite inside intracellular vesicles and can account for major fluxes of inorganic carbon to the ocean floor, although the mechanisms of calcification in coccolithophores are unclear (Berry et al. 2002). Calcification is not necessarily triggered by CO_2 limitation alone but also by availability of nutrients and the use of many different strains of coccolithophores with vastly different calcification properties. Elucidation of the molecular mechanisms for transport of Ca^{2+}, HCO_3^-, and H^+, and the interactions with nutrient acquisition, assimilation, and photosynthesis at a molecular level are required to clarify this process (Berry et al. 2002). Carbon dioxide-related effects in *Emiliana huxleyi* and the other bloom-forming coccolithophore species *Gephyrocapsa oceanica* have been observed under different physicochemical regimens (Zondervan 2007). Under high light conditions, calcification usually decreases with increasing CO_2 concentration. Depending on the nutrient status of the cells, the production of particulate organic carbon (POC) strongly increases, or decreases under elevated CO_2 concentrations. However, under low light conditions no sensitivity of calcification to CO_2 was observed, whereas POC production always strongly increases with CO_2 under nutrient-replete conditions (Zondervan 2007).

Carbonate-rich sediments represent a major $CaCO_3$ reservoir that can rapidly react to the decreasing saturation state of seawater with respect to carbonate minerals (Morse et al. 2006). Aragonite is usually the most abundant carbonate mineral in these sediments, followed by high magnesium calcite whose solubility can exceed that of aragonite. Biogenic Mg-calcites exhibit a wide range of solubilities when compared to abiotic Mg-calcites. Mg-calcite minerals will respond to rising CO_2 by sequential dissolution according to mineral stability, leading to removal of the more soluble phases until the least soluble phases remain. Laboratory experiments and observations from Bermuda and the Great Bahama Bank confirm this finding and suggest that the amount of abiotic carbonate production is likely to decline as CO_2 continues to rise (Morse et al. 2006). Foraminifera shell weight loss and bottom water carbonate content is documented (Broecker and Clark 2001). However, the initial shell thickness also varies with growth habitat, and the offset between bottom water and pore water carbonate ion concentration varies even on small space scales (Broecker and Clark 2001).

In modern seas, coralline algae produced skeletons of high Mg-calcite (>4 mol% $MgCO_3$) (Stanley et al. 2002). However, algae grown in artificial sea waters having three different Mg/Ca ratios incorporated Mg in proportion to the ambient Mg/Ca ratio. Thus, the algae calcified as if they were simply inducing precipitation from seawater through their consumption of CO_2 for photosynthesis. In the artificial sea water with a low Mg/Ca ratio of Late Cretaceous seas, algae produced low-Mg calcite (<4 mol% $MgCO_3$). Many taxa that produce high-Mg calcite today produced low Mg-calcite in Late Cretaceous seas about 65 million years ago (Stanley et al. 2002). Future anthropogenic emissions of CO_2 and the resulting ocean acidification has severe consequences for marine calcifying organisms and ecosystems (Andersson et al. 2008). Marine calcifiers depositing calcitic hard parts that contain significant concentrations of Mg-calcite and calcifying organisms living in high latitude or cold-water environments are at immediate risk because they are currently immersed in seawater that is only slightly supersaturated with respect to the carbonate phases that they secrete. Models show that high latitude ocean waters could reach aragonite undersaturation in a few decades, although at present these waters are undersaturated with Mg—calcite minerals of higher solubility than aragonite. Similarly, tropical surface seawater could become undersaturated with Mg-calcite minerals. As a result of these changes in surface seawater chemistry and further penetration of anthropogenic CO_2 into the ocean interior, magnesium content of calcitic hard parts will decrease in many ocean environments, the relative proportion of calcifiers depositing stable carbonate minerals—such as calcite and low Mg-calcite—will increase, and the mean magnesium content of carbonate sediments will decrease. Moreover, the highest latitude and deepest depth at which cold-water corals and other calcifiers currently exist will move towards lower latitudes and shallower depths, respectively (Andersson et al. 2008).

The chemical evolution of seawater over the past 600 million years is still a matter of debate (Horita et al. 2002). Based on the composition of fluid inclusions in marine halites and the mineralogy of marine evaporites, authors found two major long-term cycles in the chemistry of seawater during this period, with dramatic changes in the concentrations of Mg^{2+}, Ca^{2+}, and SO_4^-. The Mg^{2+} concentration in seawater during most of the early Paleozoic (248–543 million years ago) and Jurassic (0.6–200 mya) to Cretaceous (ended 65 mya) was as low as 30–40 mmol/kg H_2O; with maximum values recorded in the Permian (251 mya). Similarly, Ca^{2+} concentrations in seawater were two to three times greater than current values. However, SO_4^- concentrations were about a third to a fifth of modern values. The Mg^{2+}/Ca^{2+} ratio in seawater ranged from 1 to 1.5 during the early Paleozoic (248–542 mya) and Jurassic-Cretaceous (65–200 mya) to a near modern value of 5.2 during the Permian (251 mya). This change in seawater Mg^{2+}/Ca^{2+} ratio is consistent with the notion of alternating calcite-aragonite seas. Several models have been proposed to explain the chemical evolution of seawater, with mantle convection and plate activity recognized as prime movers in composition of seawater. Nevertheless, our understanding of the processes that affect composition of seawater is incomplete (Horita et al. 2002).

The decrease in the saturation state of seawater, Ω, following seawater acidification may be the major factor causing decreased calcification (Marubini et al. 2008). Inorganic precipitation of calcium carbonate was also found to be proportional to Ω, which is equal to

$$[Ca^{2+}] \times [CO_3^2]/K$$

where K is the apparent solubility constant for a particular mineral phase of $CaCO_3$. Seawater acidification was studied on the calcification and photosynthesis of the scleractinian tropical coral *Stylophora pistillata* for 8 days at pH 7.6, 8.0, or 8.2. To differentiate between the effects of the various components of the carbonate chemistry (pH, CO_3^{2-}, HCO_3^-, CO_2, Ω), duplicate tanks were maintained under similar pH but with 2 mM HCO_3^- added to the seawater. The addition of 2 mM bicarbonate significantly increased photosynthesis in *S. pistillata*, suggesting carbon-limited conditions. Conversely, photosynthesis was not affected by changes in pH or CO_2. Seawater acidification decreased coral calcification by about 0.1 mg $CaCO_3$ g daily for a decrease of 0.1 pH units. Seawater acidification affected coral calcification by decreasing the availability of the CO_3^{2-} substrate for calcification. However, the decrease in coral calcification could also be attributed to a decrease in extra- or intracellular pH or to a change in the buffering capacity of the medium, impairing supply of CO_3^{2-} from HCO_3^- (Marubini et al. 2008).

Acidification is only one of the factors affecting survival of corals About 33% of reef-building corals face elevated extinction risk from bleaching and diseases driven by elevated sea surface temperatures, increased atmospheric CO_2 concentrations, decreased ocean carbonate from acidification, increased coastal development, sedimentation from poor land-use and watershed management, sewage discharges, nutrient loading and eutrophication from agro-chemicals, coral mining, and overfishing (Carpenter et al. 2008). The Caribbean has the largest proportion of corals in high extinction risk categories whereas the western Pacific Ocean has the highest proportion of species in all categories of elevated extinction risk (Carpenter et al. 2008). For example, Caribbean coral reef communities—including fishes, corals, and macroalgae—were particularly vulnerable with overwhelming degradation as a result of agricultural land use, coastal development, overfishing, and climate change (Moya et al. 2008).

Coral reefs generally exist within a relatively narrow band of temperatures, light, and seawater aragonite saturation states (Hoegh-Guldberg 2005). The growth of coral reefs is minimal or nonexistent outside this envelope. Abnormally warm temperatures cause corals to bleach (lose their brown dinoflagellate symbionts) with almost 30% of the world's coral reefs having disappeared since the beginning of the 1980s. Increasing atmospheric CO_2 potentially lowers the aragonite saturation state of seawater, making carbonate ions less available for calcification. The synergistic interaction of elevated temperature and CO_2 is likely to produce major changes to coral reefs over the next few decades and centuries including changes in biodiversity and function of coral reefs, extinction of the reefs, and reduced prospects for industries and societies that depend on healthy coral reefs along their coastline (Hoegh-Guldberg 2005). Elevated CO_2 affects net production and calcification of a coral assemblage causing an increase in net carbon production and a decrease in calcification, lending support to the hypothesis underlying the CO_2-induced decrease in calcification of competition between photosynthesis and calcification for a limited supply of dissolved inorganic carbon (Langdon and Atkinson 2005).

Biogeochemical and physical factors that influence seawater CO_2 and air-sea CO_2 exchange was examined for one month at Hog Reef Flat, part of the rim reef of Bermuda (Bates and Samuels 2001). Factors that modified CO_2 parameters included diurnal variability, calcium carbonate production, winds, tides, platform circulation, and fluxes of offshore or onshore waters (Bates and Samuels 2001) . Growth rates of branch tips from an Atlantic scleractinian coral, *Acropora cervicornis*, were measured before, during and after exposure to elevated nitrate, phosphate, and CO_2 at 700 to 800 mg/L (Renegar and Riegl 2005). The effect of increased CO_2 on growth was greater than that of nutrient enrichment alone. High concentrations of nitrate or phosphate decreased growth rate in both the presence and absence of increased CO_2. Survival and reef-building potential of *Acropora cervicornis*, will be negatively impacted by continued coastal nitrification and projected increases in carbon dioxide (Renegar and Riegl 2005).

7.5 Literature Cited

Anderson, R.F., S. Ali, L.I. Bradtmiller, S.H.H. Nielsen, M.Q. Fleisher, B.E. Anderson, and L.H. Burckle. 2009. Wind-driven upwelling in the Southern Ocean and the deglacial rise in atmospheric CO_2, *Science*, 323, 1443–1448.

Andersson, A.J., F.T. Mackenzie, and N.R. Bates. 2008. Life on the margin: implications of ocean acidification on Mg-calcite, high latitude and cold-water marine calcifiers, *Mar. Ecol. Prog. Ser.*, 373, 265–273.

Bates, N.R., A.F. Michaels, and H. Knap. 1996. Alkalinity changes in the Sargasso Sea: geochemical evidence of calcification?, *Mar. Chem.*, 51, 347–358.

Bates, N.R. and L. Samuels. 2001. Biogeochemical and physical factors influencing seawater pCO_2 and air-sea CO_2 exchange on the Bermuda coral reef, *Limnol. Ocean.*, 46, 833–846.

Berry, L., A.R. Taylor, U. Lucken, K.P. Ryan, and C. Brownlee. 2002. Calcification and inorganic carbon acquisition in coccolithophores, *Funct. Plant Biol.*, 29, 289–299.

Brennand, H.S., N. Soars, S.A. Dworjanyn, A.R. Davis, and M. Byrne. 2010. Impact of ocean warming and ocean acidification on larval development and calcification in the sea urchin *Tripneustes gratilla*, PLoS ONE, 5(6):e11372. doi:10.1371/journal.pone.0011372.

Brewer, P.G. and J.C. Goldman. 1976. Alkalinity changes generated by phytoplankton growth, *Limnol. Ocean*, 21, 108–117.

Broecker, W. and E. Clark. 2001. An evaluation of Lohmann's foraminifera weight dissolution index, *Paleoceanography*, 16, 531–534.

Byrne, M., M. Ho, P. Selvakumaraswamy, H.D. Nguyen, S.A. Dworjanyn, and A.R. Davis. 2009. Temperature, but not pH, compromises sea urchin fertilization and early development under near-future climate change scenarios, *Proc. Roy. Soc.*, 276B, 1883–1888.

Byrne, M., M. Ho, E. Wong, N.A. Soars, P. Selvakumaraswamy, H. Shepard-Brennand, S.A. Dworjanyn, and A.R. Davis. 2010b. Unshelled abalone and corrupted urchins: development of marine calcifiers in a changing ocean, *Proc. Roy. Soc. London B*, doi.1098/rspb.2010.2404.

Byrne, M., N.A. Soars, M.A. Ho, E. Wong, D. McElroy, P. Selvakumaraswamy, S.A. Dworjanyn, and A.R. Davis. 2010. Fertilization in a suite of coastal marine invertebrates from SE Australia is robust to near-future ocean warming and acidification, *Mar. Biol.*, 157, 2061–2069.

Byrne, M. N. Soars, P. Selvakumaraswamy, S.A. Dworjanyn, and A.R. Davis. 2010a. Sea urchin fertilization in a warm, acidified and high pCO_2 ocean across a range of sperm densities, *Mar. Env. Res.*, 69, 234–239.

Carpenter, K.E., M. Abrar, G. Aeby, R.B. Aronson, S. Banks, A. Bruckner, A. Chiriboga, J. Cortes, J.C. Delbeek, L. DeVantier, G.J. Edgar, A.J. Edwards, and others. 2008. One-third of reef-building corals face elevated extinction risk from climate change and local impacts, *Science*, 321, 560–563.

Dickson, J.A.D. 2002. Fossil echinoderms as monitor of the Mg/Ca ratio of Phanerozoic oceans, *Science*, 298, 1222–1224.

Flower, B.P. 1999. Warming without high CO_2? *Nature*, 399, 313–314.

Gervais, G. and U. Riebesell. 2001. Effect of phosphorus limitation on elemental composition and stable carbon isotope fractionation in a marine diatom growing under different CO_2 concentrations, *Limnol. Ocean*, 46, 497–504.

Hoegh-Guldberg, O. 2005. Low coral cover in a high-CO_2 world, *J. Geophys. Res.* 110, C09S06, doi:10.1029/2004JC002528.

Horita, J., H. Zimmermann, and H. Holland. 2002. Chemical evolution of seawater during the Phanerozoic: implications form the record of marine evaporites, *Geochim. Cosmochim. Acta*, 66, 3733–3756.

Hutchins, D.A., M.R. Mulholland, and F. Fu. 2009. Nutrient cycles and marine microbes in a CO_2-enriched ocean, *Oceanography*, 22, 128–145.

Langdon, C. and M.J. Atkinson. 2005. Effect of elevated pCO_2 on photosynthesis and calcification of corals and interactions with seasonal change in temperature/irradiance and nutrient enrichment, *J. Geophys. Res.*, 110, C09S07, doi:10.1029/2004JC002576.

Lear, C.H., Y. Rosenthal, and N. Slowey. 2002. Benthic foraminiferal Mg/Ca-paleothermometry: a revised core-top calibration, *Geochim. Cosmochim. Acta*, 66, 3375–3387.

Lowenstein, T.K., M.N. Timofeeff, S.T. Brennan, L.A. Hardie, and R.V. Demicco. 2001. Oscillations in Phanerozoic seawater chemistry: evidence from fluid inclusions, *Science*, 295, 1086–1088.

Luthi, D., M. Le Floch, B. Bereiter, T. Blunier, J.M. Barnola, U. Siegenthaler, D. Raynaud, J. Jouzel, H. Fischer, K. Kawamura, and T.F. Stocker. 2008. High-resolution carbon dioxide concentration record 650,000–800,000 years before present, *Nature*, 453, 379–382.

Marubini, F., C. Ferrier-Pages, P. Furla, and D. Allemand. 2008. Coral calcification responds to seawater acidification: a working hypothesis towards a physiological mechanism, *Coral Reefs*, 47, 491–499.

Michaels, A.F., N.R. Bates, K.O. Buesseler, C.A. Carlson, and A.H. Knap. 1994. Carbon-cycle imbalances in the Sargasso Sea, *Nature*, 372, 537–540.

Millero, F.J., R. Woosley, B. Ditrolio, and J. Waters. 2009. Effect of ocean acidification on the speciation of metals in seawater, *Oceanography*, 22, 72–85.

Montanez, I.P. 2002. Biological skeletal carbonate records changes in major-ion chemistry of paleo-oceans, *Proc. Natl. Acad. Sci. USA*, 99, 15852–15854.

Morse, J.W., A.J. Andersson, and F.T. Mackenzie. 2006. Initial responses of carbonate-rich shelf sediments to rising atmospheric pCO_2 and "ocean acidification": role of high Mg-calcites, *Geochim. Cosmochim. Acta*, 70, 5814–5830.

Moya A., C. Ferrier-Pages, P. Furla, S. Richier, E. Tambutte, D. Allemand, and S. Tambutte. 2008. Calcification and associated physiological parameters during a stress event in the scleractinian coral *Stylophora pistillata*, *Comp. Biochem. Physiol.*, 151A, 29–36.

Pearson, P.N. and M.R. Palmer. 2000. Atmospheric carbon dioxide concentrations over the past 60 million years, *Nature*, 406, 695–699.

Pelejero, C., E. Calvo, M.T. McCulloch, J.F. Marshall, M.K. Ganan, J.M. Lough, and B.N. Opdyke. 2005. Preindustrial to modern interdecadal variability in coral reef pH, *Science*, 309, 2204–2207.

Portner, H.O. and A.P. Farrell. 2008. Physiology and climate change, *Science*, 322, 690–692.

Portner, H.O., M. Langenbuch, and B. Michaelidis. 2005. Synergistic effects of temperature extremes, hypoxia, and increases in CO_2 on marine animals: from Earth history to global change, *J. Geophys. Res.*, 110, C09S10, doi:10.1029/2004JC002561.

Renegar, D.A. and B.M. Riegl. 2005. Effect of nutrient enrichment and elevated CO_2 partial pressure on growth rate of Atlantic scleractinian coral *Acropora cervicornis*, *Mar. Ecol. Prog. Ser.*, 293, 69–76.

Retallack, G.J. 2001. A 300-million-year record of atmospheric carbon dioxide from fossil plant cuticles, *Nature*, 411, 287–290.

Ries, J.B. 2006. Mg fractionation in crustose coralline algae: geochemical, biological, and sedimentological implications of secular variation in the Mg/Ca ratio of seawater, *Geochim. Cosmochim. Acta*, 70, 891–900.

Robinson, L.F., J.F. Adkins, L.D. Keigwin, J. Southon, D.P. Fernandez, S.L. Wang, and D.S. Scheirer. 2005. Radiocarbon variability in the western North Atlantic during the last deglaciation, *Science*, 310, 1469–1473.

Rosa, R. and B.A. Seibel. 2008. Synergistic effects of climate-related variables suggest future physiological impairment in a top oceanic predator, *Proc. Natl. Acad. Sci. USA*, 105, 20776–20780.

Rothman, D.H. 2002. Atmospheric carbon dioxide levels for the last 500 million years, *Proc. Natl. Acad. Sci. USA*, 99, 4167–4171.

Sarmiento, J.L. and C. Le Quere. 1996. Oceanic carbon dioxide uptake in a model of century-scale global warming, *Science*, 274, 1346–1350.

Sluijs, A., S. Schouten, M. Pagani, M. Woltering, H. Brinkhuis, J.S.S. Damste, G.R. Dickens, M. Huber, G.J. Reichart, R. Stein, J. Matthiessen, L.J. Lourens, N. Pedentchouk, J. Backman, and K. Moran. 2006. Subtropical Arctic Ocean temperatures during the Paleocene/Eocene thermal maximum, *Nature*, 441, 610–613.

Stanley, S.M., J.B. Ries, and L.A. Hardie. 2002. Low-magnesium calcite produced by coralline algae in seawater of Late Cretaceous composition, *Proc. Natl. Acad. Sci. USA*, 99, 15323–15326.

Svensen, H., S. Planke, A. Malthe-Sorenssen, B. Jamtveit, R. Myklebust, T.R. Eidem, and S.S. Rey. 2004. Release of methane from a volcanic basin as a mechanism for initial Eocene global warming, *Nature*, 429, 542–546.

Takahashi, T., J. Olafsson, J.G. Goddard, D.W. Chipman, and S.C. Sutherland. 1993. Seasonal variation of CO_2 and nutrients in the high-latitude surface oceans: a comparative study, *Global Biogeochem. Cycles*, 7, 843–878.

Truchot, J.P. and A. Duhamel-Jouve. 1980. Oxygen and carbon dioxide in the marine intertidal environment: diurnal and tidal changes in rockpools, *Resp. Physiol.*, 39, 241–254.

UNEP (United Nations Environment Programme). 2010. Environmental consequences of ocean acidification: a threat to food security, *Available from UNEP*, P.O. Box 39552, Nairobi 00100, Kenya, 12 pp.

Velzer, J., Y. Godderis and L.M. Francois. 2000. Evidence for decoupling of atmospheric CO_2 and global climate during the Phanerozoic eon, *Nature*, 408, 698–701.

Widdicombe, S. and J.I. Spicer. 2008. Predicting the impact of ocean acidification on benthic diversity. What can animal physiology tell us?, *J. Exp. Mar. Biol. Ecol.*, 366, 187–197.

Wohlers, J., A. Engel, E. Zollner, P. Breithaupt, H.G. Hoppe, U. Sommer, and U. Riebesell. 2009. Changes in biogenic carbon flow in response to sea surface warming, *Proc. Natl. Acad. Sci. USA*, 106, 7067–7072.

Wood, H.L., J.I. Spicer, D.M. Lowe, and S. Widdicombe. 2010. Interaction of ocean acidification and temperature; the high cost of survival in the brittlestar *Ophiura ophiura*, *Mar. Biol.*, 157, 2001–2013.

Zachos, J.C., U. Rohl, S.A. Schellenberg, A. Sluijs, D.A. Hodell, D.C. Kelly, E. Thomas, M. Nicolo, I. Raffi, L.J. Lourens, H. McCarren, and D. Kroon. 2005. Rapid acidification of the ocean during the Paleocene-Eocene thermal maximum, *Science*, 308, 1611–1615.

Zhong, S. and A. Mucci. 1989. Calcite and aragonite precipitation from seawater solutions of various salinities: precipitation rates and overgrowth compositions, *Chem. Geol.*, 78, 283–299.

Zondervan, I. 2007. The effects of light, macronutrients, trace metals and CO_2 on the production of calcium carbonate and organic carbon in coccolithophores–A review, *Deep-Sea Res. II*, 54, 521–537.

CHAPTER 8

Mitigation

8.1 General

Based on current knowledge of marine impacts, Turley et al. (2010) recommend that 500 ppm atmospheric CO_2 should be avoided and that a threshold of 450 ppm CO_2 be adopted to ensure no large-scale risk to marine organisms and ecosystems. Most scientists agree that the smaller the CO_2 buildup the less the likelihood of dire impacts (Broecker 1997; Raven et al. 2005). But the demand for cheap fossil energy continues to grow. Unfortunately, no viable and acceptable option to fossil fuels has yet been devised (Broecker 1997). To curtail CO_2 emissions, it is proposed to plant forests, use alternative energy sources—such as wind, solar, and nuclear—or a combination of national and multinational cap-and-trade systems (Stavins 2009). Companies around the world would be issued rights by their governments to produce carbon, which they could buy and sell on an open market. If they wanted to produce more carbon, they could buy another company's rights. If they produced less carbon than they needed, they could buy and sell on an open market. For example, the U.S. and China—two major emitters of CO_2 (Table 8.1)—are currently involved in negotiations about climate policy. If the two nations enter into a bilateral agreement, other major nations may join. From there, developing nations could join giving them targets to reduce emissions without stifling growth. With the right incentives, developing countries will adopt less carbon-intensive growth paths (Stavins 2009). If the risk of irreversible damage arising from ocean acidification is to be avoided, particularly in the Southern Ocean, the cumulative future human derived emissions to the atmosphere must be considerably less than 900 Gt C by the year 2100, wherein 1 gigaton of carbon = 1×10^{12} kg carbon (Raven et al. 2005), but this need to be verified.

Current research goals of the U.S. National Oceanic and Research Administration include: monitoring the changing ocean chemistry and biological impacts at selected stations; developing environmental and ecological indices of ocean acidification based on sentinel organisms; developing models to predict ocean acidification impacts on biological cycling and to predict changes in the ocean carbon cycle as a function of CO_2 and climate-induced changes in temperature, ocean circulation, biogeochemistry, ecosystems and terrestrial input; increasing technology to measure carbonate chemistry; promoting satellite monitoring of reef habitats and changes in surface ocean chemistry in response to ocean acidification; and further analyzing social and economic consequences of a changing marine ecosystem (USDC 2008).

Table 8.1. The top ten CO_2-emitting countries in 2006. Total (in millions of metric tons of CO_2) and per capita emissions (in tons per capita). Modified from Stavins (2009).

Rank	Country	Total emissions	Per-capita emissions
1	China	6018	4.6
2	U.S.A.	5903	19.9
3	Russia	1704	12.0
4	India	1293	1.2
5	Japan	1247	9.8
6	Germany	858	10.4
7	Canada	614	18.8
8	U.K.	586	9.7
9	South Korea	515	10.5
10	Iran	472	7.3

8.2 Ocean Sequestration

The ocean represents the largest potential sink for anthropogenic CO_2 (Herzog 1998). Discharging CO_2 directly to the ocean would accelerate the ongoing, but slow, natural processes by which more than 90% of present day emissions are currently entering the ocean indirectly and would reduce both peak atmospheric CO_2 concentrations and their rate of increase. Options for environmentally sound, economically viable, and technically feasible ocean sequestration strategies are under consideration (Herzog 1998).

Oceanic CO_2 levels are expected to rise during the next 200 years to levels not seen for 10 to 150 million years by the uptake of atmospheric CO_2, or potentially through the disposal of waste CO_2 in the deep sea (Barry et al. 2005). As atmospheric CO_2 concentrations increase, CO_2 will diffuse into the hydrosphere (Kikkawa et al. 2003). Disposal of CO_2 at fixed locations in the ocean might give strong maximum pH reduction at depth where natural variability in pH is smaller and tolerance limits for biota probably narrower (Haugan and Drange 1996; Herzog et al. 1996; Haugan 1997).

Ocean sequestration of CO_2 as a possible measure to reduce the rate of increase of atmospheric CO_2 is proposed through 1/release of gaseous or liquid CO_2 at shallow depths of <1,000 m from the coast by pipelines, 2/discharging liquid CO_2 into intermediate ocean depths (1,000–2,000 m) from a moving ship by a towed pipe, or 3/storing liquid CO_2 in a restricted depression of the deep sea floor (>3,000 m) (Kikkawa et al. 2003). Ocean disposal of fossil fuel carbon dioxide as a solid hydrate at depths ranging from 349 to 3,627 m and from 8° to 1.6°C showed decomposition at shallow depths and chemical alterations at deeper depths (Brewer et al. 1999). These problems could be averted by first reacting waste CO_2 with water and a carbonate material, such as limestone, to form dissolved bicarbonate for release into the sea (Rau and Caldeira 2002). The addition of alkalinity to the ocean resulting from this enhanced bicarbonate production would also help to buffer ocean acidification attributable to atmospheric CO_2 (Rau and Caldeira 2002). Additional research on the suggestion of Rau and Caldeira to react waste CO_2 with carbonate seems merited (Seibel and Walsh 2002).

Stabilizing the concentration of atmospheric CO_2 may require storing captured anthropogenic CO_2 in near permanent geologic reservoirs (House et al. 2006). Injecting CO_2 into deep-sea sediments <3,000 m water depth and a few hundred meters of sediment provides permanent geologic storage, even with large geomechanical perturbations. At the high pressures and low temperatures common in deep sea sediments, CO_2 remains liquid and can be denser than the overlying pore water, causing the injected CO_2 to be gravitationally stable. Additionally, CO_2 hydrate formation will impede the flow of CO_2 and serve as a second cap. If calcareous sediments are chosen, then the dissolution of carbonate host rock by the aqueous carbon dioxide solution will increase porosity and subsequently the permeability. The total CO_2 storage capacity within the 200-mile economic zone of the U.S. coastline is capable of storing thousands of years of current U.S. emissions aver the authors (House et al. 2006). However, shallow injection near the shore may be less expensive in terms of energy and capital than deep-ocean injection (Haugan and Drange 1992).

Five major characteristics of deep-sea organisms that are relevant to CO_2 ocean sequestration include low biological activity, comparatively lengthy life span, high sensitivity to environmental disturbances, high species diversity, and low density (Shirayama 1998). Once damaged, they may become extinct or recover over a lengthy period (Shirayama 1998). Today's need for substantive CO_2 emission reductions could be satisfied more cheaply by available sequestration technologies than by an immediate transition to nuclear, wind, or solar energy (Lackner 2003). Further development of sequestration would assure plentiful, low-cost energy for the century, giving better alternatives ample time to mature (Lackner 2003). The deep-sea is a geochemical sink wherein biomass and species richness declines at depths below 1,000 m (Omori et al. 1998). Knowledge of deep-sea plankton and micronekton is insufficient to merit CO_2 disposal into the deep sea. Research is needed on the potential effects of lowered pH and raised partial pressure of CO_2 over the entire life cycle of zooplankton and micronekton under realistic hydrostatic pressures. At present, confined release of CO_2 in restricted depressions of the sea floor at depths greater than 3,000 m seems preferable to dispersion in midwater (Omori et al. 1998).

Underground storage of industrial quantities of carbon dioxide in porous and permeable reservoir rocks has been taking place for at least 5 years at the Sleipner West gas field in the North Sea (Holloway 2005). Globally, there may be sufficient underground storage capacity for CO_2 to make a significant impact on global emissions to the atmosphere. But uncertainty on actual storage capacity, safety, security of storage, and public acceptability are issues that need to be resolved (Holloway 2005).

Release of CO_2 from fossil fuel burning facilities directly into the ocean at various depths is not recommended (Ishimatsu and Kita 1999; Tamburri et al. 2000; Huesemann et al. 2002; Thistle et al. 2005). Animal responses to dissolving CO_2 hydrates on the deep-sea floor was investigated in the Monterey Canyon, California, using a submersible (Tamburri et al. 2000). Several species of invertebrates and vertebrates did not avoid rapidly-dissolving flocculent hydrates when attracted by the scent of food. Decreased pH had no apparent short-term effects; however, mobile animals appeared to suffer from respiratory distress due to increased CO_2 near the hydrates, suggesting that losses of higher organisms as a result of deep-sea CO_2 disposal may be more extensive than previously predicted from toxicological models (Tamburri et al. 2000). Results of a 30-day study on the effect of injected CO_2 on deep-sea (3,200 m) meiofaunal community in the Monterey Canyon demonstrated no change in abundance of major groups, including copepods, nematodes, polychaetes, and total meiofauna (Carman et al. 2004). Authors aver that slow decomposition rates of meiofaunal carcasses masked adverse effects of CO_2, and that longer observation periods may be necessary (Carman et al. 2004).

Chronic effects of CO_2 ocean sequestration—those directly related to the marine ecosystem—are difficult to verify experimentally or to assess using ecosystem models (Kita and Ohsumi 2004; Adams and Caldeira 2008). A suggested practical solution is field studies starting with controlled small scale CO_2 injections and eventually to a large scale CO_2 injection to determine ecosystem alterations (Kita and Ohsumi 2004). Simulations of deep ocean CO_2 injection as an alternative to atmospheric release show greater chemical and biological impact on the deep ocean as the price for having less impact on the surface ocean and climate (Caldeira and Wickett 2005). Laboratory studies indicate that acidified CO_2-containing plumes from fossil fuel plants of pH as low as 6.0 will drastically inhibit microbial nitrification, resulting in a subsequent reduction of nitrite and nitrate (Huesemann et al. 2002). Moreover, storage of anthropogenic carbon dioxide in the deep ocean is not recommended because of potential impacts on deep-living fauna (Seibel and Walsh 2001, 2003). Low metabolic rates of deep-sea fauna are correlated with low capacities of pH buffering and low concentrations of ion-transport proteins. Accordingly, changes in seawater CO_2 may lead to large cellular CO_2 and pH changes. Oxygen transport proteins of deep-sea fauna are also sensitive to changes in pH. Acidosis leads to metabolic suppression, reduced protein synthesis, respiratory stress, reduced metabolic scope, and ultimately death, Much additional research needs to be conducted before injection of CO_2 into the abyss is used as a means of controlling atmospheric CO_2 levels (Seibel and Walsh 2001, 2003). The effects of deep-sea carbon dioxide injection on fauna were evaluated during a field experiment in 3,600 m depth off California, in which liquid CO_2 was released on the sea floor (Barry et al. 2004). Exposure to the dissolution plume from the liquid CO_2 resulted in high rates of mortality for flagellates, amoebas, and nematodes that frequent sediments in close proximity to sites of CO_2 release. Seawater pH reductions of 0.5 to 1.0 pH units near CO_2 release sites are fatal to nearby infaunal deep-sea communities (Barry et al. 2004, 2005). Mortality associated with exposure to low pH can be avoided by properly dispersing the CO_2 and keeping the plume off the seabed (Caulfield et al. 1997). Reduced zooplankton mortality from ocean CO_2 disposal occurs when the number of diffuser points per point source is increased from 2 to 12 (Adams et al. 1997). However, some species of amphipods and eels are unusually abundant in the vicinity of deep-sea (1,300 m) hydrothermal vents near Hawaii that discharge CO_2 plumes with a mean pH of 6.7 (Vetter and Smith 2005). Results suggest that at least some species of scavenging amphipods and eels can escape relatively concentrated CO_2 plumes and revive following 60-min exposure to such plumes. In addition, results suggest that these species of scavenging amphipods and eels can detect and avoid intoxicating levels of CO_2 resulting from deep ocean injection of CO_2 and that specialized components of the deep sea fauna are well-adapted to exploit carrion accumulating on the periphery of injection sites (Vetter and Smith 2005).

8.3 Declining Water Quality

Attempts to mitigate CO_2 emissions suggest that reef managers and coastal resource policies must first reduce the influence of declining water quality, coastal pollution, and overexploitation of key functional groups such as herbivores (as quoted in Hoegh-Guldberg et al. 2007). New techniques should be developed for the mass culture of corals from fragments and spat in an attempt to assist local restoration, or the culture of resistant varieties of key organisms. Attempts should be made to facilitate grazing by sea urchins and fish, especially parrotfish, provided that water quality and other factors are not limiting. Authors state that restriction of atmospheric CO_2 concentration to <500 ppm will benefit coral reefs and the millions of people who depend on them, but are not hopeful that they will be implemented given scientific reticence and the inherently conservative nature of consensus seeking (as quoted in Hoegh-Guldberg et al. 2007).

8.4 Reduction in Emissions from Airliners

The U.S. Federal Aviation Administration and the European air-traffic organization Sesar are joined in an initiative called the Atlantic Interoperability Initiative to Reduce Emissions (Michaels 2010). They found that reducing engine thrust when possible and flying at optimal altitudes can result in fuel savings of 1.2% for over-oceanic flights and 20% for landing approaches. However, this involves greater communication between airlines, airports, and air-traffic controllers for agreement on common goals (Michaels 2010).

8.5 Increasing International Cooperation

Modern climate change is dominated by human influences, which are now large enough to exceed the bounds of natural variability (Karl and Trenberth 2003). There is still considerable uncertainty about the rates of change that can be expected, but anthropogenic climate change is now likely to continue for many centuries. Climate change is unlikely to be adequately addressed without greatly improved international cooperation and action (Karl and Trenberth 2003). The ultimate goal of the UN Framework Convention on Climate Change is to achieve stabilization of greenhouse gas concentrations at a level that would prevent anthropogenic interference with the climate system (Wigley et al. 1996). With the concentration targets yet to be determined, Working Group I of the Intergovernmental Panel on Climate Change (IPCC) developed a set of illustrative pathways for stabilizing the atmospheric CO_2 concentration at 350, 450, 550, 650, and 750 ppm over the next few hundred years. Unfortunately, these models depended on a transition from the current heavy dependence on fossil fuels. Rising sea-levels and their economic consequences seem to be key in regard to the optimal timing of IPCC mitigation measures (Wigley et al. 1996).

The Kyoto Protocol was introduced in 1997 and entered into force in 2005, with the aim of stabilization of greenhouse gas concentrations in the atmosphere that would prevent "dangerous anthropogenic interference" with the climate system (Hardman-Mountford et al. 2008). The Protocol requires countries to reduce greenhouse gas emissions below specified levels, with some signatory nations setting more stringent targets. Part of the post-Kyoto carbon accounting requires that changes in natural sinks of CO_2 be tracked and if possible predicted. Automated measurement of CO_2 in the Southern Ocean, Atlantic Ocean, and northwestern European shelf seas is now operational and could be expanded (Hardman-Mountford et al. 2008). "Dangerous anthropogenic interference" can be viewed from a variety of perspectives (O'Neill and Oppenheimer 2002). Current agreement on this term includes large-scale eradication of coral reef systems, CO_2 stabilization at 450 ppm, sustained warming in excess of 2°C above 1990 global temperatures (all with a high probability of occurring within 2 to 200 years), and disintegration of the West Antarctic ice sheet. The latter event is a process expected to occur over the next 5 to 50 centuries but would raise sea level 4 to 6 m. All targets are marked by a high degree of uncertainty in the working of the carbon cycle (O'Neill and Oppenheimer 2002). The biggest polluters—the United States and China, which account for about a quarter of the global population and nearly half its toxic emissions—are not bound by the agreement (Fears 2010). Japan—which is a signatory to the protocol—said that they would no longer lower its emissions to the detriment of its economy while pollution engines in China and the United States were at full throttle and would withdraw from the protocol when its current phase expires in 2012 (Fears 2010).

Arctic countries are encouraged to ratify the United Nations proposal (the Gothenburg Protocol) to abate acidification, eutrophication, and ground level ozone (Calder et al. 2006). As of July 2006, Denmark, Finland, Norway, Sweden, and the United States have signed and ratified the Protocol; Canada has signed but not yet ratified the Protocol; and Iceland and the Russian Federation have neither signed nor ratified the Protocol. Additional cooperation is merited on sources of Arctic acidification from oil and gas activities, smelters, and from short-term episodic events such as volcanic eruptions and forest fires (Calder et al. 2006).

The longer we put off serious action, the more aggressive our future efforts will need to be as carbon-spewing capital assets continue to accumulate (Stavins 2009). Plants built today will determine emissions for a generation. Most steel plants, as one example—where plant lifetimes typically exceed 25 years are now less than 10 years old. If the consequences of continued inaction are correct, the time has come for meaningful and sensible action (Stavins 2009). Even global climate model scenarios that assume significantly reduced greenhouse gas emissions through technological advancements and societal change indicate that warming alone will likely continue through the year 2300 (Ruhl et al. 2008).

Participants at the *Second Symposium on the Ocean in a High-CO_2 World* (Orr et al. 2009) stressed the importance of improving international coordination to facilitate agreements on protocols, methods, and data reporting in order to optimize limited resources by greater sharing of materials, facilities, expertise and data. Despite major uncertainties, the participants urged the research community to expedite an understanding of individual organisms' responses to provide meaningful predictions of the effects of oceanic acidification on food webs, fisheries, marine ecosystems, coastal erosion, and tourism. Finally, managers and policy makers need to easily understand information on indicators of change and thresholds beyond which marine ecosystems will not recover (Orr et al. 2009).

The United Nations Framework Convention of Climate Change plans to establish a global mechanism for curbing deforestation (Eilperin 2010). Deforestation currently accounts for about 15% of the global annual greenhouse gas emissions. Brazil, Indonesia, and Papua New Guinea are among the nations where forests are being cut to make way for additional cattle grazing areas, and for the growing of soybeans and palms. Norway has pledged more than 1 billion dollars (US) between now and 2012 as part of its pact with Brazil, Indonesia, and Guyana, and the USA has promised another billion as part of any broad international agreement. Although at least a dozen countries have promised to help pay to preserve forests in the short term, the effort will probably fall short of what is needed to reduce deforestation in half by 2020, namely 25 billion (US) per year (Eilperin 2010).

8.6 Develop Alternative Technologies

The climatic impact of fossil energy can be reduced by separating the resulting carbon and sequestering it away from the atmosphere (Parson and Keith 1998). Air capture is an industrial process for recapturing CO_2 from ambient air and is one of an emerging set of technologies for CO_2 removal that includes geological storage of biotic carbon and acceleration of geochemical weathering (Keith 2009). Although this new technology is comparatively expensive, air capture allows the application of industrial economies of scale to small and mobile emission sources and a partial decoupling of carbon capture for the energy infrastructure, advantages that may compensate for the intrinsic difficulty of capturing carbon from air (Keith 2009). Despite environmental concerns, more than 30 traditional coal plants have been built or are under construction in the United States since 2008 with a combined annual generation of 125 million tons of greenhouse gases to the atmosphere annually (Brown 2010). The new plants do not capture carbon dioxide, despite the infusion of \$3.4 billion in stimulus spending and \$687 million spent by the U.S. Department of Energy on clean-coal programs. Coal-burning utilities will remain the largest industrial source of climate-changing gases for years to come until implementation of cost-effective carbon-neutralizing technologies projected to begin in about 20 years (Brown 2010).

Mexico, with the 14th largest economy in the world, contributes between 1.5 and 3% of CO_2 emissions (Booth 2010). Mexico is a major producer of cement wherein the cement kilns burn dirty, cheap fuels such as oil sludge. Cement makers are now exploring ways to burn sewage and other fuels in redesigned, cleaner kilns (Booth 2010).

8.7 Environmental Modification

Mesoscale experiments have demonstrated that iron additions enhance phytoplankton growth and reduce surface CO_2 levels in high-nutrient, low-chlorophyll regions of the world oceans (Lam et al. 2001). The addition of iron specifically stimulates organic but not inorganic carbon production in the Subarctic Pacific Ocean. Authors conclude that iron fertilization may be particularly effective in reducing CO_2 in surface waters by stimulating primary production but not calcium carbonate precipitation which augments CO_2 (Lam et al. 2001).

Controls on water acidification and de-oxygenation in an estuarine waterway located in eastern Australia became necessary as a consequence of land drainage since the early 1900s (Lin et al. 2004). Acidic flows of <pH 4.5 occurred intermittently, with acidity due mainly to oxidation of Fe^{2+} and subsequent hydrolysis of Fe^{3+} (Lin et al. 2004). This appears to be a common problem in mining communities (Eisler 2004).

Reversal of the growth of atmospheric methane and other trace gases would avert anthropogenic interference with global climate (Hansen and Sato 2004). Reductions in trace gas emissions may allow stabilization of anthropogenic CO_2 at an achievable level of anthropogenic CO_2 emissions. A decline of non-CO_2 forcings allows climate forcing to be stabilized with a significantly higher transient level of CO_2 emissions. Increased emissions of CO_2, N_2O, and CH_4 are expected in response to global warming, but these emissions are small compared to anthropogenic emissions; however, they tend to aggravate stabilization of atmospheric composition (Hansen and Sato 2004).

8.8 Legislation

The Center for Biological Diversity—based mainly on information presented in previous chapters of this work—has requested that the U.S. Environmental Protection Agency (EPA) publish revised water quality criteria taking into account new information about ocean acidification, thus marking the first step toward a national approach of preventing carbon dioxide pollution from degrading oceanic water quality (Sakashita 2007). This request was made pursuant to section 304 of the Clean Water Act, wherein EPA has listed pH as a pollutant in its regulations and developed pH water quality in 1976 (namely, for open ocean waters where the depth is substantially greater than the euphotic zone, the pH should not be changed more than 0.2 units outside the naturally occurring variation or in any case outside the range of 6.5 to 8.5). New information on ocean acidification, however, has rendered the existing water quality criteria for pH outdated and inadequate.

Specifically, the petition requests that EPA 1/revise pH criteria to reflect new ocean acidification science and adopt a criterion prohibiting any measurable change in pH of marine waters, and 2/publish information concerning acidification to guide states in monitoring and preventing harmful ocean acidification. This petition is for a non-discretionary action under the Clean Water Act and therefor the agency is required to respond and the action is enforceable (Sakashita 2007).

8.9 Literature Cited

Adams, E.E. and K. Caldeira. 2008. Ocean storage of CO_2, *Elements*, 4, 319–324.

Adams, E.E., J.A. Caulfield, H.J. Herzog, and D.I. Auerbach. 1997. Impacts of reduced pH from ocean CO_2 disposal: sensitivity of zooplankton mortality to model parameters, *Waste Mange.*, 17, 375–380.

Barry, J.P., K.R. Buck, C.F. Lovera, L. Kuhnz, P.J. Whaling, E.T. Peltzer, P. Walz, and P.G. Brewer. 2004. Effects of direct ocean CO_2 injection on deep-sea meiofauna, *J. Ocean*, 60, 759–766.

Barry, J.P., K.R. Buck, C. Lovera, L. Kuhnz, and P.J. Whaling. 2005. Utility of deep sea CO_2 release experiments in understanding the biology of a high-CO_2 ocean: effects of hypercapnia on deep sea meiofauna, *J. Geophys. Res.*, 110, C09S12, doi:10.1029/2004JC002629.

Booth, W. 2010. Mexico aims to be a leader in emissions reduction, *Washington Post* (newspaper), Nov. 29, 2010, A9.

Brewer, P.G., G. Friederich, E.T. Peltzer, and F.M. Orr Jr. 1999. Direct experiments on the ocean disposal of fossil fuel CO_2, *Science*, 284, 943–945.

Broecker, W.S. 1997. Thermohaline circulation, the Achilles heel of our climate system: will man-made CO_2 upset the current balance?, *Science*, 278, 1582–1588.

Brown, M. 2010. More coal plants are under construction. Old-style power source on rise despite environmental concerns, *Washington Post* (newspaper), Aug. 23, A3.

Caldeira K. and M.E. Wickett. 2005. Ocean model predictions of chemistry changes from carbon dioxide emissions to the atmosphere and ocean, *J. Geophys. Res.*, 110, C09S04, doi:10.1029/2004JC002671.

Calder, J., Y. Tsaturov, P. Dovle, R. Shearer, M. Olsen, O. Mahonen, H. Jensson, G. Futsaeter, C. de Wit, and J.I. Solbakken. 2006. Arctic pollution 2006. Acidification and Arctic haze, *Arctic Monitoring and Assessment Programme, P.O. Box 8100 Dep, N-0032, Oslo, Norway*, 28 pp.

Carman, K.R., D. Thistle, J.W. Fleeger, and J.P. Barry. 2004. Influence of introduced CO_2 on deep-sea metazoan meiofauna, *J. Ocean*, 60, 767–772.

Caulfield, J.A., D.I. Auerbach, E.E. Adams, and H.J. Herzog. 1997. Near field impacts of reduced pH from ocean CO_2 disposal, *Energy Convers. Manage.*, 38, Suppl.1, S343–S348.

Eilperin, J. 2010. U.N. nears agreement to curb destruction of world's rain forests, *Washington Post* (newspaper), Dec. 8, 2010, A11.

Eisler R. 2004. *Biogeochemical, Health, and Ecotoxicological Perspectives on Gold and Gold Mining*, CRC Press, Boca Raton, Florida, 355 pp.

Fears, D. 2010. What was behind Japan's surprise stand? *Washington Post* (newspaper), Dec. 8, 2010, A11.

Hansen, J. and M. Sato. 2004. Greenhouse gas growth rates, *Proc. Natl. Acad. Sci USA*, 101, 16109–16114.

Hardman-Mountford, N.J., G. Moore, D.C.E. Bakker, A.J. Watson, U. Schuster, R. Barciela, A. Hines, G. Moncoiffe, J. Brown, S. Dye, J. Blackford, P.J. Somerfield, J. Holt, D.J. Hydes, and J. Aiken. 2008. An operational monitoring system to provide indicators of CO_2-related variables in the ocean, *ICES J. Mar. Sci.*, 65, 1498–1503.

Haugan, P.M. 1997. Impacts on the marine environment from direct and indirect ocean storage of CO_2, *Waste Manage.*, 17, 323–327.

Haugan, P.M. and H. Drange. 1992. Sequestration of CO_2 in the deep ocean by shallow injection, *Nature*, 357, 318–320.

Haugan, P.M. and H. Drange. 1996. Effects of CO_2 on the ocean environment, *Energy Convers. Mgmt.*, 37, 1019–1022.

Herzog, H.J. 1998. Ocean sequestration of CO_2–an overview, *Proc. Fourth Inter. Conf. Greenhouse Gas Control Technol., Aug. 30-Sep. 2, 1998, Interlaken, Switzerland*, 7 pp.

Herzog, H.J., E.E. Adams, D. Auerbach, and J. Caulfield. 1996. Environmental impacts of ocean disposal of CO_2, *Energy Convers. Mgmt.*, 37, 999–1005.

Hoegh-Guldberg, O., P.J. Mumby, A.J. Hooten, R.S. Steneck, P. Greenfield, E. Gomez, C.D. Harvell, P.F. Sale, A.J. Edwards, K. Caldeira, N. Knowlton, C.M. Eakin, R. Iglesias-Prieto, N. Muthiga, R.H. Bradbury, A. Dubi. and M.E. Hatziolos. 2007. Coral reefs under rapid climate change and ocean acidification, *Science*, 318, 1737–1742.

Holloway, S. 2005. Underground sequestration of carbon dioxide—a viable greenhouse gas mitigation option, *Energy*, 30, 2318–2333.

House, K.Z., D.P. Schrag, C.F. Harvey, and K.S. Lackner. 2006. Permanent carbon dioxide storage in deep-sea sediments, *Proc. Natl. Acad. Sci. USA*, 103, 12291–12295.

Huesemann, M.H., A.D. Skillman. and E.A. Crecelius. 2002. The inhibition of marine nitrification by ocean disposal of carbon dioxide, *Mar. Pollut. Bull.*, 44, 142–148.

Ishimatsu, A. and J. Kita. 1999. Effects of environmental hypercapnia on fish, *Japan. J. Ichthyol.*, 46, 1–13.

Karl, T.R. and K.E. Trenberth. 2003. Modern global climate change, *Science*, 302, 1719–1723.

Keith, D. 2009. Why capture CO_2 from the atmosphere?, *Science*, 325, 1654–1655.

Kikkawa, T., A. Ishimatsu, and J. Kita. 2003. Acute CO_2 tolerance during the early developmental stages of four marine teleosts, *Environ. Toxicol.*, 18, 375–382.

Kita, J. and T. Ohsumi. 2004. Perspectives on biological research for CO_2 ocean sequestration, *J. Ocean*, 60, 695–703.

Lackner, K.S. 2003. A guide to CO_2 sequestration, *Science*, 300, 1677–1678.

Lam, P.B., P.D. Tortell, and F.M.M. Morel, 2001. Differential effects of iron additions on organic and inorganic carbon production by phytoplankton, *Limnol. Ocean.*, 46, 1199–1202.

Lin, C., M. Wood, P. Haskins, T. Ryffel, and J. Lin. 2004. Controls on water acidification and de-oxygenation in an estuarine waterway, eastern Australia, *Estuar. Coast. Shelf Sci.*, 61, 55–63.

Michaels, D. 2010. Airlines find ways to trim fuel use, *Wall Street J.*, March 10, 2010, A18.

Omori, M., C.P. Norman, and T. Ikeda. 1998. Oceanic disposal of CO_2: potential effects on deep-sea plankton and micronekton. A review, *Plankton Biol. Ecol.*, 45, 87–99.

O'Neill, B.C. and M. Oppenheimer. 2002. Dangerous climate impacts and the Kyoto Protocol, *Science*, 296, 1971–1972.

Orr, J.C., K. Caldeira, V. Fabry, J.P. Gattuso, P. Haugan, P. Lehodey, S. Pantoja, H.O. Portner, U. Riebesell, T. Trull, E. Urban, M. Hood, and W. Broadgate. 2009. Research priorities for understanding ocean acidification. Summary from the Second Symposium on the Ocean in a High-CO_2 World, *Oceanography*, 22, 182–189.

Parson, E.A. and D.W. Keith. 1998. Climate change: fossil fuels without CO_2 emissions, *Science*, 382, 1053–1054.

Rau, G.H. and K. Caldeira. 2002. Minimizing effects of CO_2 storage in oceans, *Science*, 295, 275–276.

Raven, J., K. Caldeira, H. Elderfield, O. Hoegh-Guldberg, P. Liss, U. Riebesell, J. Shepard, C. Turley, and A. Watson. 2005. Ocean acidification due to increasing atmospheric carbon dioxide, *Policy doc. 12/05, The Royal Society, 6–9 Carlton House Terrace, London SW15AG*, 57 pp.

Ruhl, H.A., J.A. Ellena, and K.L. Smith Jr. 2008. Connections between climate, food limitation, and carbon cycling in abyssal sediment communities, *Proc. Natl. Acad. Sci. USA*, 195, 17006–17011.

Sakashita, M. (Staff attorney). 2007. Petition for revised pH water quality criteria under section 304 of the Clean Water Act. 33 U.S.C. § 1314, to address ocean acidification, Petition submitted to U.S. Environmental Agency by Center for Biological Diversity.1095 Market St., Suite 511, San Francisco, California, 21 pp.

Seibel, B.A. and P.J. Walsh. 2001. Potential impacts of CO_2 injection on deep-sea biota, *Science*, 294, 319–320.

Seibel, B.A. and P.J. Walsh. 2002. Response (to Rau and Caldeira, 2002), *Science*, 295, 276–277.

Seibel, B.A. and P.J. Walsh. 2003. Biological impacts of deep-sea carbon dioxide injection inferred from indices of physiological performance, *J. Exp. Biol.*, 206, 641–650.

Shirayama, Y. 1998. Biodiversity and biological impact of ocean disposal of carbon dioxide, *Waste Manage.*, 17, 381–384.

Stavins, R.N. 2009. Can countries cut carbon emissions without hurting economic growth? *Wall Street J.* (newspaper), September 21, 2009, R1–R3.

Tamburri, M.N., E.T. Peltzer, G.E. Friederich, I. Aya, K. Yamane, and P.G. Brewer. 2000. A field study of the effects of CO_2 ocean disposal on mobile deep-sea animals, *Mar. Chem.*, 72, 95–101.

Thistle, D., K.R. Carman, L. Sedlacek, P.G. Brewer, J.W. Fleeger, and J.P. Barry. 2005. Deep-ocean, sediment-dwelling animals are sensitive to sequestered carbon dioxide, *Mar. Ecol. Prog. Ser.*, 289, 1–4.

Turley, C., M. Eby, A.J. Ridgwell, D.N. Schmidt, H.S. Findlay, C. Brownlee, U. Riebesell, V.J. Fabry, R.A. Feely, and J.P. Gattuso. 2010. The societal challenge of ocean acidification, *Mar. Pollut. Bull.*, 60, 787–792.

USDC (United States Department of Commerce), National Oceanic and Atmospheric Administration (NOAA). 2008. Ocean acidification, *State of the Science Fact Sheet*, 2 pp.

Vetter, E.W. and C.R. Smith. 2005. Insights into the ecological effects of deep ocean CO_2 enrichment: the impacts of natural CO_2 venting at Loihi seamount on deep sea scavengers, *J. Geophys. Res.*, 110, C09S13, doi:10.1029/2004JC002617.

Wigley, T.M.L., R. Richels, and J.A. Edmonds. 1996. Economic and environmental choices in the stabilization of atmospheric CO_2 concentrations, *Nature*, 379, 240–243.

CHAPTER 9

Concluding Remarks

Ocean chemistry—including carbonate content and pH—has remained constant for hundreds of thousands of years, with typical surface ocean variation between ice ages and warming periods of less than 0.2 pH units. Since the late 1790s, however, anthropogenic combustion of fossil fuels has resulted in an unprecedented increase of atmospheric CO_2 of about 100 ppm to 350 ppm CO_2. The oceans have taken up about 30% of the CO_2 from the atmosphere where it forms carbonic acid and this is projected to lower surface ocean water pH an estimated 0.3 units by the year 2100. Such a rapid change in ocean pH has very likely not happened in 65 million years. Dissolution of oceanic carbonate sediments will eventually restore oceanic pH to its pre-industrial level, a process estimated to take more than a million years. During the last glacial period, about 8,000 years ago, global atmospheric CO_2 concentrations were reduced by 80 ppm; however, dissolution of foraminiferan tests in the western equatorial Atlantic Ocean was documented at about the same time. The reasons for these phenomena are unknown, again illustrating the uncertainties associated with oceanic acidification.

Major current sources of carbon dioxide include fossil fuel combustion and deforestation while major sinks include the atmosphere and the oceans. During glacial periods the oceans are sinks for atmospheric CO_2 and during glacial-interglacial transitions the oceans are a source of CO_2 to the atmosphere, although controlling mechanisms remain unresolved. Since the 1790s the oceans have absorbed about 127 billion tons of carbon as CO_2 from the atmosphere, or about 33% of the anthropogenic carbon emitted. Increasing CO_2 concentrations in seawater results in formation of carbonic acid, which causes acidification. Carbonic acid and water form bicarbonate ions, reducing carbonic acid and calcite (a form of calcium carbonate). Elevated CO_2 in seawater also results in an increase in bicarbonate. Models predict that future anthropogenic CO_2 inputs to the atmosphere will continue to be sequestered in the ocean. The role of calcium carbonate in carbon sequestration merits additional research on the mechanistic controls of these seemingly coupled processes. Other sources of CO_2 to the ocean include volcanos, deep-sea hydrothermal vents, and riverine discharges of acidified freshwater. Phytoplankton, especially coccolithophores, play a major role in carbon uptake and export, as do calcifiers such as diatoms, reef-building corals, bivalve and gastropod molluscs, and many species of crustaceans and echinoderms. At present, most available evidence suggests that atmospheric concentrations of CO_2 will continue to increase during the next 100 years and that oceanic pH will continue to decrease; however, the rates at which these processes occur are the subject of much discussion and uncertainty.

The ocean carbon cycle is continually modified by the increases in atmospheric CO_2 due to fossil fuel combustion. The global coastal ocean today is a net sink of atmospheric CO_2. Additional chemical information is needed on the role of carbonates in calcification and dissolution chemistry and on acidification processes. Excess organic carbon formed at higher CO_2 levels could increase the expansion of "dead zones", but the mechanisms of the process require clarification. Information is also needed on various physical processes including interhemispheric carbon transport, sinking rates of phytodetritus, vertical transport of particulate matter, oxygen consumption by sediments and sediment pore waters, and improved measurements of saturation horizons for calcite and aragonite. Acidification mode of action varies greatly among different species of marine flora and fauna. Among photosynthetic flora, the role of pH, bicarbonate, CO_2, carbonate, and inorganic and organic carbon compounds influence species succession and distribution, but species differ in uptake rate, nutrient status, photosynthetic rates, and carbon fixation. In other taxonomic groups, major research areas requiring additional research effort include calcification mechanisms, nutrient uptake, carbon concentrating mechanisms, and dissolution chemistry in echinoderms, corals, molluscs, and crustaceans; acidosis, acid-base transfers, and significance of compensating bicarbonate exchange mechanisms in annelids, arthropods, elasmobranchs and bony fishes; and the role of myoglobin, hypometabolism, and protein phosphorylation in marine reptiles, birds, and mammals.

By the year 2100, atmospheric carbon dioxide concentration is expected to exceed 500 ppm, surface water pH is projected to drop to about 7.8, and aragonite and calcite undersaturation may occur in the Southern Ocean, the Arctic, and parts of the Bering Sea and Chukchi Sea. During this period global temperatures may rise, resulting in ice melt and subsequent flooding of coastal areas. However, the rate of increase in global temperature and that of sea level rise are both uncertain. Results of laboratory studies indicate that most species of marine biota can tolerate a pH decrease of 0.5 units for extended periods and that 500 ppm of atmospheric CO_2 will benefit most seagrasses and selected phytoplankton. However, many species will suffer under this regimen. Among sea urchins, high mortality and test dissolution were found at 500 mg/L CO_2 and pH 7.8; other adverse effects of this high CO_2-low pH regimen include reduced urchin sperm motility, reduced sperm swimming speed, low fertilization success, inhibited larval development, and reduced growth. Corals are particularly vulnerable to acidification, especially in regard to fertilization, settlement success, and calcification. But the current contradictory reports about likely effects of atmospheric CO_2 addition to the ocean and subsequent oceanic acidification on corals, and other calcifiers, indicate that predictions need to be reexamined. Among molluscs, bivalve larvae were the most sensitive stage at 500 mg/L CO_2 and pH 7.8, but resistance varied among species. Cuttlefish and other decapod molluscs were comparatively resistant to elevated CO_2 and reduced pH. Shrimps avoided low pH environments. Vertebrates appeared more resistant than plants and invertebrates, but data were insufficient to draw conclusions. Should atmospheric CO_2 exceed 650 ppm CO_2 (projected to occur by the year 2200), laboratory studies now conclude that shellfish populations will be depleted and coral reefs will begin to dissolve.

In general, results of field investigations tend to support those of laboratory studies for a high CO_2 (500 ppm), low pH (7.8) regimen. Unfortunately, there exist few field data of sufficient duration, resolution, and accuracy on the acidification rate and the factors governing its variability. Results of current field studies suggest that locales with the highest potential risk of future acidification include the Arctic Ocean, Southern Ocean, North Atlantic Ocean, North Sea, and the central North Pacific Ocean. Arctic Ocean surface waters, for example, are vulnerable to aragonite undersaturation, and will become more sensitive as atmospheric CO_2 concentrations exceed 450 ppm. Aragonite undersaturation together with decreasing sea ice cover in the summer, and decreasing carbonate ion concentrations, will adversely affect many species of Arctic Ocean calcifiers. A similar case is made for the Southern Ocean. There, aragonite undersaturation as a result of atmospheric CO_2 levels greater than 450 ppm will adversely impact calcification in pteropod veligers and foraminiferans. In the North Atlantic Ocean, thermohaline circulation weakens with global warming and collapses at high levels of CO_2. In the North Sea, jellyfish abundance has increased since 1958, especially since 1970, with abundance positively correlated with decreasing pH. In the central North Pacific Ocean, there was a significant trend over a period of 20 years in decreasing surface pH that correlated with rate of acidification expected for equilibration with the atmosphere. Many other oceanic locations are now experiencing high CO_2 and low pH.

Reconciliation of the findings from field investigations with those of laboratory studies is extremely difficult owing to the large number of physical, chemical, and biological variables known to affect ocean surface pH and atmospheric CO_2. For example, natural variations in seawater pH and atmospheric carbon dioxide levels over the millennia are documented that preclude anthropogenic input. Further, the mineral composition and crystallization rate of $CaCO_3$ precipitates from the ocean are significantly modified by Mg/Ca ratios of the ambient seawater as well as by orthophosphate ions, sulfate concentration of the fluids, temperature, and salinity. Future anthropogenic emissions of CO_2 and the resulting ocean acidification has severe consequences for marine calcifying organisms and ecosystems. Calcifiers depositing calcitic hard parts that contain elevated concentrations of magnesium calcite and those living in cold-water environments are at immediate risk because they are immersed in seawater undersaturated with aragonite and calcite. Changing ocean pH levels also radically affects the abundance of various metal salts with as yet unknown effects on the entire marine food chain. Seawater chemistry merits additional research effort to resolve these issues. Information on organism adaptation or acclimatization to increasing ocean acidification and their impact on species survival is woefully inadequate. No less important are clarification of the mechanisms of action of acidification. Continued field monitoring of resources at risk is recommended as are laboratory studies fueled by field results. Other stressors that would probably exacerbate acidification effects on biota include pollution, sea-level rise, climate change, and habitat destruction from mining and overfishing.

All concerned scientists now agree that efforts to curtail anthropogenic carbon dioxide emissions to the atmosphere should include extensive planting of forests and increasing use of alternative energy sources. Most scientists endorse: increased use of satellites to monitor the changing ocean chemistry and biotic composition at selected stations; wider use of sentinel organisms as indicators of acidification; increased efforts to develop mathematical models to predict rate of acidification, impacts on biota, and ocean circulation; and further analyzing social and economic consequences of changing marine ecosystems. Many scientists now disagree that oceanic sequestration of liquid or solid carbon dioxide in selected geologic reservoirs—including the abyss—is desirable. The scientific community, in general, endorses the development of alternative technologies to fossil energy, including the recapture of CO_2 from the ambient air, development of carbon neutralizing technologies, environmental modification through controls on water acidification, and new technologies that reduces CO_2, methane, and other greenhouse gases from coal and other fossil fuels.

Since impending acidification of the oceans is a global problem affecting all nations, international cooperation between the United States and the People's Republic of China—the two largest emitters of anthropogenic CO_2 to the atmosphere—is mandatory. The cooperation of India, Russia, and Japan is also needed, and these together with the U.S. and China could provide recommendations to the United Nations that would be agreeable to less developed nations. Ultimately, the United Nations should be empowered to set oceanic pH criteria and to act as a clearinghouse for information on the latest CO_2-containing procedures. Failure to act positively within the next few decades would probably result in much property loss via coastal flooding as well as destruction of valuable marine resources.

General Index

A

Abalone 124, 125, 200, 201
Acetazolamide 62, 77, 81
Acid-base 52, 53, 59, 80–86, 101, 133, 136, 137, 145, 146, 148, 200, 229
Acid-base equilibria 52
Acid-base imbalance 53, 86
Acid-base regulation 59, 80, 83, 84, 200
Acid-iron 127
Acid rain 5
Acid sulfate 132, 143
Acidic deposition 6, 169
Acidic gases 6
Acidic sulphate soils 28
Acidification 2, 3, 5, 6, 16, 21, 24, 28, 32, 36, 40, 41, 50, 51, 53, 56, 58, 59, 63, 67, 72–75, 78, 81, 99–126, 128–130, 132–135, 137–145, 148, 149, 164, 165, 169, 172–174, 176–178, 180, 184–187, 193, 199–202, 206–208, 213–215, 217, 220, 221, 223, 224, 228–232
Acidification rates 3, 32, 35, 56, 59, 65, 68, 71, 73, 74, 102, 103, 105, 115, 119, 122, 170, 175
Acidified ecosystems 6
Acidified lowlands 40
Acidosis 59, 60, 80, 81, 83–87, 101, 122, 134–137, 145–147, 215, 229
Africa 11
Agricultural land use 208
Agro-chemicals 208
Airborne fraction 23, 24
Alanine 121
Alaska 182
Alaskan crab fisheries 184
Aleutian Basin 172

Algae 33, 35, 38, 51, 55, 60, 62, 65, 70–72, 74–76, 101–104, 108, 109, 114–117, 124, 165, 171, 206
Algal calcification 51, 70
Alkaline pH 52, 65, 80, 134, 147
Alkalinity 24, 27, 32, 37, 41, 54, 72, 114, 116, 167, 170, 171, 173, 184, 195, 203, 204, 215
Alkenone isotopes 16
Alkenones 15
Alternative energy sources 213, 232
Alternative technologies 222, 232
Aluminum 147–149, 203
Amino acid 127
Ammonia 83, 110, 113
Ammonium 60, 110, 113, 116, 126, 182, 187
Amoebas 218
Amphipods 129, 130, 218
Annelids 79, 126, 229
Antarctic 14, 87, 132, 147, 185, 198, 220
Antarctic bottom water 185
Antarctic Ocean 185
Antarctic Polar Front 198
Antarctica 12–15, 26, 36, 54, 185, 186, 198
Anthropogenic CO_2 3, 11, 12, 22–26, 33, 41, 50, 51, 53, 55, 56, 100, 105, 112, 115, 166, 167, 175, 176, 178, 179, 185, 186, 206, 214, 215, 223, 228, 229, 232
Arabian Sea 12, 166, 176
Aragonite 4, 5, 17, 37, 50, 51, 53–59, 70–73, 75, 78, 99, 100, 115, 117, 118, 122, 124, 144, 164–166, 170, 174, 176–178, 184, 186, 195, 196, 205–208, 229–231

Aragonite saturation 51, 55, 70, 71, 83,
    115, 117, 118, 166, 174, 176, 177,
    184, 208
Aragonite undersaturation 56, 99, 164,
    166, 186, 206, 231
Arbroath, Scotland 62
Arctic 12, 16, 40, 56, 99, 124, 131, 165,
    166, 220, 230, 231
Arctic Ocean 40, 56, 99, 131, 165, 230,
    231
Arctic Sea 12
Arizona 6
Arterial blood 82, 84, 146
Arterial pH 82, 83, 87, 143, 146
Arthropods 80, 121, 124, 127, 229
Asparagine 127
Aspartic acid 71, 127
Asteroids 200, 247
Atlantic Interoperability Initiative to
    Reduce Emissions 219
Atlantic Ocean 12, 22, 24, 31, 32, 49,
    50, 55, 58, 65, 66, 102, 164, 166, 167,
    169, 178, 179, 197, 198, 202, 220,
    228, 230, 231
Atmospheric carbon dioxide 1, 3, 4, 5,
    11, 14, 22, 26, 99, 107, 173, 183, 187,
    193, 196, 198, 230, 231
Atmospheric CO_2 3, 4, 11–17, 21–24,
    27–34, 49–51, 54–57, 60, 61, 67,
    69–71, 99, 104, 105, 109, 110,
    113, 114, 117, 119, 128, 135, 164,
    166, 167, 169, 172, 173, 177, 178,
    184–187, 194, 196–199, 202, 208,
    213–215, 217–219, 228–231
Atmospheric oxygen 51
ATP 68, 80, 121, 180
Australia 28, 33, 35, 39, 76, 170, 200,
    223
Austria 6

B

Bacteria 34, 57, 79, 80, 101, 182
Baltic Sea 111, 171, 178
Barbados 73
Barnacles 130, 131
Barrier reef 33, 35, 36, 76, 170
Bay of Bengal 176

Belgium 171
Bering Sea 56, 172, 184, 230
Bermuda 30, 36, 58, 96, 168, 169, 172,
    195, 205, 209
Beryllium 203, 204
Biocalcification 122
Bicarbonate 3, 22, 52, 59, 60, 64, 66, 68,
    73, 76, 77, 80, 82, 83, 87, 100, 104,
    107–109, 114, 122, 127, 134, 136,
    141, 146, 207, 215, 229
Bioerosion 36, 38, 116
Biogenic calcification 32, 35, 53, 68, 75
Biogenic carbon 29, 69, 199
Biogenic carbonates 36
Biogenic opal 198
Biogenic production 49
Biological pump 57, 172
Bioturbation 138, 183
Birds 88, 140, 148, 188, 230
Bivalves 78, 123
Bleaching 71, 103, 119, 120, 208
Blood oxygen 84, 101, 143
Blood pigments 101
Blood plasma 86, 146
$B(OH)_3$ 196
$B(OH)_4^-$ 196
Borneo 173
Boron 14, 15, 21, 196, 198
Boron isotope composition 196, 198
Boron isotopes 14, 15, 21
Brazil 221
British Columbia 182
Brittle star 37, 46
Brunei estuary 173
Bryozoans 139

C

^{14}C-aspartic acid 71
^{45}Ca 71
Ca^{2+} 53, 70, 72, 81, 86, 137, 193, 205, 207
Ca-ATPase 68
$CaCl_2$ 70
$CaCO_3$ 4, 13, 22, 26, 27, 29, 30, 32–38,
    40–42, 49, 50, 52–55, 71, 72, 74, 75,
    86, 100, 104, 116, 122, 165, 168, 171,
    173, 178, 186, 205, 207, 208, 231

Cadmium 147–149
Calcareous algae 35, 70
Calcareous plankton 184
Calcareous skeletal structures 53
Calcification 1, 4, 17, 22, 23, 27, 29,
    31–33, 35, 36, 49, 51, 53, 56, 58, 59,
    65–77, 100–107, 114–120, 122–125,
    130, 134, 135, 138, 165, 169–171,
    173–176, 181, 184, 187, 195, 200,
    201, 205, 207–209, 229–231
Calcifiers 73
Calcifying flora 53
Calcite 4, 16, 22, 40, 51, 54–58, 65,
    66, 72, 99, 100, 102, 104, 105, 115,
    118, 129, 164, 165, 169, 170, 173,
    175–177, 181, 186, 187, 196, 197,
    205–207, 229–231
Calcite-aragonite 207
Calcite compensation depth 197
Calcite saturation 176, 197
Calcium 4, 5, 16, 17, 22, 27, 32, 35, 37,
    38, 51–55, 58, 60, 68, 70–72, 75, 78,
    85, 99, 100, 104, 105, 107, 108, 114,
    122, 123, 129, 131, 132, 134, 139,
    148, 149, 169, 170, 186, 196, 201,
    207, 209, 223, 229
Calcium carbonate 4, 5, 22, 27, 32, 35,
    37, 38, 51–55, 58, 68, 70, 72, 75, 78,
    100, 105, 107, 108, 114, 122, 129,
    139, 169, 170, 186, 196, 201, 207,
    209, 223, 229
Calcium carbonate dissolution 114,
    129
Calcium carbonate saturation 5, 35, 51,
    53, 54, 70, 107, 201
Calcium channels 71
California 38, 216, 218
Cambrian 17, 195
Canada 6, 178, 214, 220
Canary Islands 65
Cap-and-trade 213
Carbon 1–5, 11–17, 21–27, 29–42, 49–
    55, 57–69, 71–80, 84, 99, 101–103,
    105–117, 124, 128, 129, 132, 133,
    136, 137, 144, 146, 147, 166–170,
    172–183, 185–187, 193–200, 203,
    205, 207–209, 213–218, 220–223,
    228–232

Carbon assimilation number 179
Carbon chemistry 3
Carbon-concentrating mechanisms 60,
    67, 72, 74, 229
Carbon cycling 29, 32, 34, 194, 195
Carbon dioxide 1–5, 11, 14–17, 22,
    24–26, 30, 32, 42, 50, 52, 53, 58,
    59, 66, 71, 74, 75, 78, 79, 84, 99,
    101, 107, 110, 112, 116, 117, 128,
    132, 133, 136, 137, 144, 146, 147,
    168, 173, 177, 183, 187, 193–196,
    198–200, 205, 209, 215–218, 222,
    223, 228, 230–232
Carbon dioxide fixation rate 66
Carbon dynamics 195
Carbon fixation 31, 33, 66, 75, 79, 105,
    112, 169, 174, 229
Carbon flux 13, 36, 37, 57, 69, 166, 183
Carbon isotope 15, 32, 196, 205
Carbon sequestration 26, 27, 57, 109,
    229
Carbon production 30, 31, 36, 62, 106,
    169, 175, 208, 222
Carbon transport 38, 57, 72, 75, 195,
    229
Carbonate 3–5, 15–17, 22, 23, 26–28, 32,
    35, 37–39, 49–58, 61, 68–70, 72–76,
    78, 81, 85, 86, 99, 100, 102, 104, 105,
    107, 108, 114–119, 122, 124, 129,
    134, 137, 139, 165, 167, 169–173,
    176, 183, 186, 187, 193–197, 201,
    203, 205–209, 214, 215, 223, 228,
    229, 231
Carbonate chemistry 35, 56, 61, 69, 70,
    105, 207, 214
Carbonate compensation depth 15
Carbonate dissolution 26, 114, 129,
    173, 176
Carbonate ions 52, 53, 70, 74, 208
Carbonate mineral saturation state 173
Carbonate precipitation 27, 223
Carbonate production 27, 32, 38, 39,
    49, 55, 75, 85, 105, 114, 206, 209
Carbonate saturation 5, 35, 49–51,
    53–55, 70, 99, 100, 107, 116, 165,
    201
Carbonate systems 49

Carbonic acid 3, 21, 22, 50, 52, 78, 114, 228, 229
Carbonic anhydrase 62–65, 67, 68, 72, 75, 77–79, 81, 85, 86
Cardiac failure 143, 146
Cardiac output 82, 84, 143, 146
Cardiorespiratory responses 82, 143, 146
Caribbean 114, 116, 167, 168, 173, 208
Cascades 6
Catecholamines 85
CCM 60, 74
Cement 3, 25, 57, 116, 222
Cement production 3, 25, 57
Cenozoic 196
Center for Biological Diversity 223
Cerium 203, 204
Cetaceans 88
CH_4 223
Chaetognaths 140
China 24, 179, 213, 214, 220, 232
Chlorofluorocarbons 2
Chlorophyll 29–31, 65, 110, 169, 179, 222
Chlorophytes 72, 100
Chukchi Sea 56, 230
Citrate synthase 146
Cl^- 73, 80, 81, 83, 85–87, 145
Clams 78, 79, 120, 123, 126
Cl^-/HCO_3^- 73, 83, 87
Cl^-/OH^- 73, 80
Clean Water Act 223, 224
Climate change 12, 26, 35, 40, 41, 50, 51, 70, 121, 133, 166, 167, 179, 199–201, 208, 219, 221, 232
Climate events 197
Climate policy 213
Cnidarians 77
C:N:P 110
Coastal acidification 40
Coastal development 208
Coastal erosion 221
Coastal nitrification 209
Coastal pollution 218
Coastal protection 56, 119
Coastal reefs 5, 113

CO_2 1–4, 11–17, 21–42, 49–87, 99–117, 119, 121–149, 164–173, 175–180, 182–187, 194, 196–209, 213–223, 228–232
CO_2-concentrating mechanisms 60, 67, 72, 74, 229
CO_2 fixation 31, 33, 66, 75, 79, 105, 112, 169, 174, 229
CO_2 hydrates 216
CO_2 ocean sequestration 216, 217
CO_2 sources and sinks 25
CO_3^{2-} 52, 53
Cobalt 203
Coccolithophore calcification 65, 195
Coccolithophore fluxes 174
Coccolithophores 22, 32, 38, 61, 66, 68, 75, 104–106, 118, 165, 168, 169, 175, 205, 229
Coccoliths 30, 54, 104, 181
Coelenterates 69, 100, 113
Commercial airliners 24
Conifer forests 6
Copepods 37, 40, 127–129, 131, 217
Copper 6, 109, 129, 133, 145, 203, 204
Copper smelters 6
Coral calcification 53, 69, 73, 76, 117, 169, 170, 207, 208
Coral reef 5, 33, 34, 36, 51, 69, 70, 76, 77, 100, 113, 114, 116, 117, 142, 169, 184, 199, 208, 220
Corals 4, 5, 17, 35, 38, 51, 53, 55, 69–77, 100, 101, 114, 115, 118, 119, 137, 141, 165, 170, 176, 184, 198, 206, 208, 218, 229, 230
Coralline algae 33, 51, 71, 72, 75, 101–103, 108, 117, 171, 206
Coralline macroalgae 4
Corallite 118
Cortisol 85, 143
Crabs 59, 81, 133, 134, 177
Cretaceous 17, 108, 196, 206, 207
Crustaceans 34, 37, 59, 70, 80, 100, 101, 132, 134, 148, 229
Crustose algae 33
Crustose coralline algae 33, 71, 102, 117, 171
Cryptophores 182

Cu 145, 204
Cuttlefish 125, 230
Cyanobacteria 66, 76, 111, 171

D

DDT 147
Dead zones 50, 229
Decarbonizing 24
Deep-sea CO_2 disposal 217
Deep Western Boundary Current 167
Deforestation 21, 221, 228
Deglaciation 13, 198
Denmark 6, 220
Devil's Hole 172
Devonian 17
Diatoms 29, 30, 34, 63, 72, 110, 129,
    168, 169, 172, 182, 229
DIC 39, 63–65
3(3,4-dichlorophenyl)-1,1-dimethyl
    urea 68
4,4'-diisothiocyanato-stilbene-2,2'-
    disulfonic acid 73
Dimethylsulphide 30
Dinoflagellates 63, 71–74, 77, 78, 129,
    208
Dissolution 4, 13, 26, 27, 39–42, 50–56,
    60, 75, 78, 79, 100, 103, 104, 114,
    116, 117, 120–124, 129, 130, 134,
    135, 137, 165, 171–174, 176, 181,
    186, 197, 205, 215, 218, 228–230
Dissolution event 13
Dissolution of $CaCO_3$ 41, 50, 55
Dissolution process 56
Dissolution rate 26, 27, 176
Dissolved CO_2 77, 85, 99, 142, 205
Dissolved free carbon dioxide 3
Dissolved inorganic carbon 16, 22–24,
    30, 33, 41, 55, 61–64, 72–74, 76, 77,
    109, 168, 176, 199, 209
Dissolved organic phosphorus 77
Dogfish 82, 83, 140, 143, 146
Dover Strait 37
Dysprosium 203, 204

E

Earth 11, 13, 41, 102
East Greenland Current 174

Echinoderms 34, 37, 70, 100, 101, 121,
    135, 137, 139, 165, 194, 229
Echinoids 38, 136, 194, 200
Eels 84, 87, 218
Eifuku volcano 177
Elasmobranchs 82, 83, 140, 146, 229
Embryonic development 120, 121, 129,
    130, 132
English Channel 37, 110
Environmental modification 222, 232
Eocene 15, 16, 196, 197
EPA 223, 224
Epinephrine 85, 143
Equator 29, 57, 166, 167, 179, 181
Erbium 203, 204
Euphausids 127, 131, 247
Euphotic zone 30, 36, 37, 180, 181, 183,
    198, 223
Eurasian Rivers 40
Europe 6, 24
Europium 203, 204
Eutrophication 183, 200, 208, 220
Evaporites 195, 196, 207
Evapotranspiration 40
Extracellular CO_2 59
Extracellular HCO_3^- 145
Extracellular pH 80, 127, 147
Eyjafjallajokull volcano 39

F

Fe^{2+} 28, 223
Fe^{3+} 28, 223
Fecal pellet 36, 123
Finfish 140
Finland 6, 220
Fish 38, 140, 143, 148, 149
Fishes 83, 140
Fishery Resources 140
Flagellates 218
Florida 75, 114
Foraminifera 13, 15, 21, 54, 67, 72, 106,
    167, 187, 198, 206
Foraminiferans 4, 35, 38, 53, 75, 106,
    165, 187, 195, 231
Fossil fuel combustion 3, 5, 6, 21, 24,
    33, 49, 99, 228, 229

Fossil fuels 1–3, 5, 23–25, 52, 173, 193, 213, 219, 228, 232
France 6
French Polynesia 33, 34, 36
Fringing reef 33, 36

G

gapdh gene 130
Gadolinium 203, 204
Galapagos Rift 79
Gallium 203
Gametogenesis 117
Gastropods 70, 100, 124
Gene expression 100, 136
Genomics 100, 138
Germany 6, 214
Gibralter Strait 63
Glacial cycles 198
Glacial-interglacial 13, 14, 21, 54, 107, 228
Glacial periods 21, 32, 228
Glacial time 167
Global climate 12, 15, 17, 26, 30, 221, 223
Global warming 16, 50, 167, 199, 202, 223, 231
Glyceraldehyde-3-phosphate dehydrogenase 130
Glutamic acid 127
Glycogen 87, 127
Gothenburg Protocol 220
Great Bahama Bank 205
Great Barrier Reef 33, 35, 76, 170
Greenhouse gases 1, 2, 14, 15, 222, 232
Greenland 174, 198
Greenland Sea 174
Gulf of Maine 174, 175
Guyana 221

H

H⁺ 28, 50, 52, 64, 77, 78, 80, 81, 83, 84, 86, 204
H⁺-ATPase 78, 83
H⁺-ATPases 80
Hawaii 26, 39, 42, 76, 117, 129, 180, 218

HCO₃⁻ 22, 50, 52, 63–65, 73, 74, 78, 79, 80, 83–87, 141, 145, 205, 207
H₂CO₃ 21, 52
Heart rate 141, 143, 146, 149
Heat shock protein 130
Hematocrit 143, 144, 146
Hemolymph 60, 121, 122, 133
Heptachlor 147
Hog Reef Flat 209
Holmium 203, 204
Holocene 14, 21, 167, 187, 198
Honolulu 76
Hsp70 gene 130
Hydrogen 3, 4, 23, 52, 100, 165, 177
Hydrogen sulfide 177
Hydrothermal vents 39, 79, 129, 218, 229
Hypercarbia 82
Hypercapnia 59, 60, 80, 82–85, 87, 101, 111, 115, 122, 123, 133, 134, 140, 143–145, 149, 201
Hypometabolism 88, 89, 230
Hypothermia 88
Hypoxic zones 53, 99

I

Ice ages 196, 228
Ice cores 12–14, 26, 32, 198
Iceland 39, 202, 220
Iceland Sea 202
India 214, 232
Indian Ocean 12, 22, 36, 41, 175, 176
Indium 203, 204
Indonesia 221
Industrial Revolution 26, 54
Inorganic aragonite 184
Inorganic calcite 175
Inorganic carbon 16, 21–24, 27, 29, 30, 33, 35–41, 51, 55, 60–68, 71–77, 79, 80, 102, 103, 106, 108, 109, 129, 167–169, 172, 174–176, 186, 199, 205, 209, 222
Inorganic cementation 116
Insecticides 147
Intergovernmental Panel on Climate Change 35, 70, 167, 219

Intertidal plants 64
Intracellular pH 73, 80, 85, 116, 146, 147, 208
Invertebrates 2, 29, 34, 53, 101, 112, 121, 137, 188, 199, 216, 230
Ion-regulatory epithelia 59
Iran 214
Iron 28, 34, 58, 112, 127, 171, 203, 222
Irradiance 35, 64, 65, 74, 76, 103, 108, 119
Ischia Island 176
Italy 176

J

Japan 177, 179, 214, 220, 232
Japan/East Sea 179
Jellyfish 179, 231
Jurassic 207
Jurassic-Cretaceous 207

K

K^+ 81
Kaneohe Bay 117
Key Largo 75
Kenyan reefs 38
Krill 132
K'sp 53
Kyoto Protocol 219

L

Labrador Sea 177
Lactate 143, 147
Lactic dehydrogenase 146
Lanthanum 203, 204
LCDW 57
Lead 22, 27, 33, 36, 61, 71, 80, 81, 109, 125, 126, 129, 139, 148, 149, 171, 181, 203, 204, 217
Lecithotrophy 138
Legislation 2, 223
Limestone 215
Lipids 65
Liquid CO_2 215, 218
Lizard Island 33
Lobsters 135
Loihi seamount 39, 129

Lower Circumpolar Deep Water 57
Lutetium 203, 204
Lysocline 165, 181, 186, 197
Lysosomal membrane 122

M

Macroalgae 4, 33, 60, 62–64, 107, 108, 208
Macrophytes 60, 74, 148, 149
Magnesium 16, 17, 39, 50, 51, 54, 57, 72, 85, 102, 132, 134, 135, 165, 205, 206, 231
Magnesium calcite 51, 54, 72, 102, 165, 205, 231
Malathion 147
Mammals 88, 140, 149, 188, 230
Mangrove ecosystems 100
Mangroves 114
Mantle convection 207
Mariana arc 177
Marine ecosystems 11, 26, 32, 54, 56, 99, 115, 128, 136, 187, 200, 221, 232
Mass extinction 115
Meiofauna 217
Mercury 109, 148, 149
Mesozoic 17
Mesozooplankton 181
Metabolic CO_2 35, 72, 86
Metal complex 52
Metal smelting 6
Metal sulfides 28
Metal uptake 108, 125, 133
Metazoan macrofauna 183
Methane 2, 14, 16, 197, 223, 232
Methoxychlor 147
Methylmercury 148, 149
Mexico 6, 178, 222
Mg 23, 31, 32, 41, 54, 61–63, 65–67, 75, 84–87, 100, 103–107, 109–112, 116–119, 121–125, 127, 128, 130–140, 142–145, 147, 168, 173, 175, 179, 182, 184, 193–196, 205, 206, 208, 230, 231
Mg^{2+} 56, 86, 137, 207
Mg/Ca 118, 194, 206, 231
Mg^{2+}/Ca^{2+} 207
Mg-calcite 100, 173, 205, 206

MgCO$_3$ 206
MgSO$_4$ 195
Microbial community 31, 182
Microbial food web 61
Microbial nitrification 217
Microbial pathogens 134
Micronekton 216
Middle Atlantic Bight 41
Migration 37, 143, 202
Mining 208, 223, 232
Miocene 15, 197, 198
Miocene Climactic Optimum 197
Mitigation 2, 40, 113, 199, 213, 219
Molluscs 34–37, 53, 59, 72, 78, 100, 101,
    120–122, 124, 125, 128, 141, 148,
    177, 229, 230
Molybdenum 112, 125, 126, 134, 139
Monoamino dicarboxylic acids 127
Monterey Bay 38
Monterey Canyon 216, 217
Moorea, French Polynesia 33
mRNA 138
Mussels 120, 122, 123, 177
Myoglobin 88, 230
Mysidaceans 247

N

N$_2$ 66, 67, 203
N$_2$-fixation 29, 67, 111, 112
N$_2$O 223
N:P ratios 66, 110
Na$^+$ 73, 80, 81, 83–87, 145
Na$^+$/H$^+$ 80, 81, 83, 84
Na$_2$CO$_3$ 70
NaHCO$_3$ 73
Na$^+$/H$^+$/HCO$_3^-$/Cl$^-$ 81
Na$^+$/K$^+$-ATPase 80, 145
Na$^+$/NH$_4^+$ 83
Nanophytoplankton 172
Nannoplankton 31, 194
Nekton 41
Nematodes 217, 218
Nemerteans 126
Neodymium 203, 204
Nephrocalcinosis 144
Netherlands 6
Neutral red retention 122

New York Bight 127
New Zealand 144
NH$_3$ 24, 83
NH$_4^+$ 83, 204
Nickel 203, 204
Nitrate 105, 107, 109, 110, 113, 116, 126,
    138, 169, 180, 182, 187, 203, 217
Nitric 5
Nitrite 110, 113, 126, 180, 187, 217
Nitrogen 5, 6, 24, 29, 34, 66, 67, 77,
    106, 109–112, 122, 171, 180, 203,
    204
Nitrogen fixation 29, 67, 111, 112
Nitrogen flux 180
Nitrogen oxides 6
Nitrous oxide 2
NO$_3^-$ 204
Nordic Sea 12
Norepinephrine 85, 143
North America 6, 16, 24, 40, 178
North Atlantic Ocean 12, 24, 50, 164,
    167, 169, 178, 179, 198, 202, 230,
    231
North Atlantic Oscillation 168
North Atlantic spring bloom 168
North Pacific Ocean 42, 50, 54, 56
North Pole 196
North Sea 31, 40, 65, 171, 178, 179, 216,
    230, 231
Norway 6, 220, 221
Norwegian-Greenland Seas 174
Norwegian Sea 174
Nutrient cycling 113, 138, 182
Nutrient dynamics 180, 188
Nutrient uptake 77, 106, 109, 229
Nutrients 29, 34, 42, 49, 67, 76, 77, 109,
    110, 113, 119, 174, 184, 191, 202, 205

O

^{18}O 194
Ocean acidification 1–3, 16, 24, 32, 41,
    50, 51, 53, 56, 59, 73, 75, 100–102,
    112–116, 118, 119, 121–124, 126,
    128, 129, 134, 135, 138–140, 142,
    164, 165, 172–174, 177, 178,
    184–187, 193, 199, 200, 202, 206,
    213–215, 223, 224, 231

Ocean atmosphere carbon exchange study 167
Ocean circulation 1, 164, 214, 232
Ocean pH 2, 3, 5, 11, 27, 50, 51, 54, 112, 115, 177, 180, 228, 231
Ocean sequestration 2, 214–217
Ocean thermohaline circulation 199, 231
Ocean warming 41, 199–201
Oceanic acidification 21, 28, 58, 72, 106, 115, 178, 193, 200, 221, 228, 230
Oceanic alkalinity 32
Oceanic carbon dioxide 42
Oceanic pH 3, 4, 11, 21, 123, 147, 228, 229, 232
OH⁻ 83, 203
Oil sludge 222
Okhotsk Sea 184
Oligocene 15
Opal 169, 198
Orbital eccentricity 14
Ordovician 17
Organic calcite 175
Organic carbon 3, 22, 27, 29–32, 34, 36, 37, 40, 41, 50, 58, 61, 62, 66, 68, 105, 109, 114, 169, 170, 172, 174, 175, 178, 180, 182, 183, 186, 187, 196, 199, 205, 229
Organic pollution 27
Orthophosphate 193, 231
Otoliths 144, 145
Overfishing 140, 200, 208, 232
Oxygen 1, 14, 50, 51, 53, 58, 59, 62, 76, 80, 82–84, 87, 88, 101, 107, 122, 125, 131–134, 137, 143, 144, 146–148, 171, 180, 196, 200, 202, 203, 217, 229
Oxygen debt 202
Oxygen isotopes 196
Oysters 120–126
Ozone 6, 220

P

Pacific Ocean 5, 12, 15, 22, 37, 42, 49, 50, 54, 56, 66, 79, 105, 117, 172, 179–184, 198, 202, 208, 222, 231

Paleocene 15, 16, 196, 197
Paleocene-Eocene 16, 197
Paleocene-Eocene thermal maximum 16
Paleozoic 17, 207
Papua New Guinea 221
Parrotfish 218
Particulate inorganic carbon 61, 68, 102, 106, 169
Particulate organic carbon 29–31, 40, 58, 61, 68, 169, 180, 183, 186, 205
Pearl River estuary 27
Permian 195, 207
Persian Gulf 166
PFD 64, 65
pH 2–6, 11, 15–17, 21, 22, 24, 26–29, 33, 39–41, 50–54, 56, 59, 60, 62–66, 68, 69, 71, 73–78, 80–87, 99–103, 107–110, 112–149, 165–169, 171, 173, 176–181, 187, 193, 196–204, 207, 208, 215–218, 223, 224, 228–232
pH criteria 224, 232
Phagocytosis 123
Phenoloxidase 134
Phosdrin 147
Phosphate 64, 87, 105, 109, 126, 130, 138, 182, 209
Phosphorus 34, 66, 67, 77, 110–112, 180, 187, 203, 205
Photon fluence density 64
Photosynthesis 29, 32, 35, 36, 60–71, 73–79, 101, 103, 104, 106, 107, 110, 112, 114, 119, 169, 171, 175, 178, 187, 202, 205–207, 209
Photosynthetic flora 60, 102
Photosynthetic rate 32, 66, 74, 77, 103
Phycocyanin-rich 111
Phycoerythrin-rich 111
Phytodetritus 57, 229
Phytoplankton 1, 13, 29–32, 34, 41, 58, 60–62, 65, 68, 72, 101, 102, 109–111, 128, 138, 168, 169, 171, 172, 174, 182, 202–204, 222, 229
PIC 68, 69
PIC/POC 69
Picoplankton 31
Piscivores 148, 149

Plankton 30, 32, 34, 41, 105, 169, 184, 203, 216
Plasma chloride 144
Plasma sodium 144, 145
Plate activity 207
Pleistocene 15, 21, 60, 117
POC 68, 69, 105, 186, 205
Polar melting 15
Pollution 27, 200, 218, 220, 223, 232
Polychaetes 217
Polysaccharide particles 58
Pore water pH 50, 129
Praseodymium 203, 204
Prawns 132
Precambrian 195
Primary production 13, 29, 31, 33, 35, 37, 102, 104, 105, 169, 170, 174, 203, 223
Promethium 203, 204
Protein degradation 122
Protein phosphorylation 87, 230
Protists 34, 35, 69, 113, 182
Prymnesiophytes 182
Pteropods 4, 54, 72, 165, 186, 195
Pycnocline 40, 41

Q

Quaternary 198

R

Radiocarbon 198
Rays 140
Reactive carbon 11
Red Sea 140, 166, 184
Redox processes 203
Remineralization 27, 30, 50, 69, 180, 183, 195
Reptiles 87, 140, 147, 148, 230
Respiratory acidosis 83, 137, 145
Reproduction 101, 121, 128, 132, 140, 143, 146, 202
Reunion Island 36
Rhodophytes 64, 65, 100
Ribulose-1,5-bis-phosphate carboxylase/oxygenase carboxylation 64
Riverine discharges 40, 178, 229

Riverine input 171
Rockies 6
Rocky Mouth Creek 28
Ross Sea 36, 164, 186
Russia 214, 220, 232
Russian Federation 220

S

Samarium 203, 204
Sargasso Sea 170, 195
Satellite monitoring 214
Saturation horizon 53, 55, 58, 176, 197
Scallops 123, 126
Scheldt estuary 171
Scheldt river 171
Scleractinian corals 17, 55, 70, 72, 76, 115, 119
Scotland 62
Sea cucumbers 137
Seagrasses 63, 64, 107, 108, 230
Sea ice 165, 166, 231
Sea-level rise 200, 232
Sea of Okhotsk 179
Sea surface temperatures 41, 60, 187, 196, 208
Sea urchins 36, 38, 100, 118, 128, 135–137, 139, 218, 230
Seawater chemistry 1, 17, 24, 53, 55, 100, 122, 170, 194, 195, 200, 206, 231
Seawater pH 3, 15, 16, 21, 26, 63, 64, 99, 103, 113, 116, 122, 124, 142, 147, 168, 180, 193, 218, 231
Second Symposium on the Ocean in a High-CO_2 World 221
Sediment nutrient flux 112, 126
Sediment oxidation 40
Sediment oxygen 14, 58
Sediment pore waters 21, 229
Sediment traps 26, 27, 172, 174, 181, 186, 187, 195
Sedimentation 32, 40, 61, 165, 166, 208
Selenium 34
Sentinel organisms 214, 232
Sequestration 2, 26, 27, 31, 57, 109, 133, 134, 214–217, 229, 232
Sesar 219

Settlement success 116, 230
Sewage 208, 222
Sharks 140
Shellfish 2, 5, 28, 99, 123, 184, 230
Shrimps 132, 230
Sierra Nevadas 6
Silicate 110, 113, 138, 168, 182, 193, 197
Silicon 198, 203
Silurian 195
Silver 125, 126
Skeletal density 76, 170
Skeletogenesis 136, 138, 201
Sleipner West gas field 216
SO$_4^-$ 207
South Atlantic Ocean 31, 58, 167, 197, 202
South Carolina 133, 146
South China Sea 179
South Korea 214
Southern hemisphere deglaciation 13
Southern Ocean 5, 12, 13, 22, 23, 56, 99, 132, 175, 179, 185–187, 198, 202, 213, 220, 230, 231
Spain 63
Sperm agglutination 136
Sperm flagellar motility 137
Sperm motility 121, 133, 134, 137, 139, 143, 230
Sponges 182
Squids 202
Sr^{2+} 56
Sr/Ca 118
St. John 115
Steel plants 221
Stony corals 100
Strontium 57, 109, 132
Subarctic Pacific Ocean 54, 55, 172, 222
Subtropical convergence 166, 175
Subtropical front 37
Subtropical gyres 31, 41, 66
Succinate 121
Sulfate 40, 57, 116, 132, 143, 193, 231
Sulfate emissions 40
Sulfide 28, 40, 79, 177
Sulfidic sediments 39
Sulfur 5, 6, 24
Sulfur dioxide 6

Sulfuric 5, 28, 84
Sulfuric acid 28, 84
Superoxide anion 123
Surface ocean pH 123
Surface water pH 4, 56, 99, 114, 136, 230
Surface water carbonate 49, 56
Sweden 6, 220
Switzerland 6
Symbionts 35, 68, 69, 77, 79, 80, 116, 117, 208
Symbiotic algae 55, 114–116

T

Tatoosh Island 187
TEP 61
Terbium 203, 204
Termination V 14
Terra Nova Bay 36, 186
Tertiary period 60
Thermal stratification 30
Thorium 58
Thulium 203, 204
Tourism 56, 100, 113, 119, 221
Trace metals 101
Transparent expolymer particle 61
Triassic 17
Tunicates 34, 38
Turonian 16, 17
Turtles 87, 147, 148
Tyrrhenian Sea 176

U

Underground storage 216
U.K. 168, 214
UN Framework Convention on Climate Change 219
United Nations 220, 221, 232
United States 6, 113, 174, 220, 222, 232
Uranium 203, 204
Urchin 38, 135–139, 176, 201, 230
U.S. 6, 113, 115, 213, 214, 216, 219, 222, 232
U.S.A. 214
U.S. Department of Energy 222
U.S. Environmental Protection Agency 223

U.S. Federal Aviation Administration
223
U.S. National Oceanic and Research
Administration 214

V

Veligers 121, 124, 186, 231
Vertical migration 37, 202
Vertical transport 58, 229
Viral lysing 34
Virgin Islands 115
Volcanic CO_2 vents 176
Volcanism 16, 39, 195
Volcano Islands 177
Volunteer observing ships 168, 173
Vostok ice core 13

W

Waikiki Aquarium 76
Washington 124, 187

Water quality 218, 223
Weddell deep water 185
Weddell Sea 185, 202
West Antarctic ice sheet 220
Western Pacific Ocean 179, 198, 208
Willapa Bay 124

Y

Yellow Sea 179
Yonge reef 35
Ytterbium 203, 204
Yttrium 109, 203, 204

Z

Zinc 109, 125, 126, 134, 139, 203, 204
Zooplankton 41, 72, 127, 216, 218
Zooxanthellae 71, 73–75, 77

Species Index

ALGAE AND OTHER PHOTOSYNTHETIC FLORA

Algae 33, 35, 38, 51, 55, 60, 62, 65, 70–72, 74–76, 101–104, 108, 109, 114–117, 124, 165, 171, 206

Chlamydomonas reinhardtii 110

Chrysophytes 34

Aureococcus anophagefferens 34

Coccolithophores 22, 32, 38, 61, 66, 68, 75, 104–106, 118, 165, 168, 169, 175, 205, 229

Calcidiscus leptoporus 61
Coccolithus pelagicus 61, 106, 174
Emiliana huxleyi 30–32, 40, 61, 62, 65, 68, 104–106, 174, 205
Gephyrocapsa oceanica 68, 104, 205

Coralline algae 33, 51, 71, 72, 75, 101–103, 108, 117, 171, 206

Amphiroa anceps 103
Amphiroa foliacea 103
Lithophyllum cabiochae 103
Porolithon onkodes 103

Crustose coralline algae 33, 71, 102, 117, 171

Neogoniolithon sp. 194

Cyanobacteria 66, 67, 76, 111, 171

Aphanizomenon sp. 171
Crocosphaera sp. 112
Nodularia spumigena 111, 171
Prochlorococcus sp. 111
Synechococcus sp. 111
Trichodesmium sp. 66, 67, 111

Diatoms 29, 30, 34, 63, 72, 110, 129, 168, 169, 172, 182, 229

Nitzschia spp. 110
Phaeocystis spp. 182
Phaeodactylum tricornutum 63, 204
Skeletonema costatum 62, 110, 204
Thalassiosira pseudonana 109
Thalassiosira rotula 57
Thalassiosira weissflogii 63

Dinoflagellates 63, 71–74, 77, 110, 129, 218

Amphidinium carterae 63
Heterocapsa oceanica 63
Phaeocystis globosa 62
Symbiodinium sp. 78

Foraminifera 13, 15, 21, 64, 67, 72, 106, 167, 187, 198, 206

Amphisorus lobifera 67
Amphistegina hemprichii 67
Globigerina bulloides 187
Globigerinoides sacculifer 106, 167
Neogloboquadrina dutertrei 167

Macroalgae 4, 33, 60, 62–64, 107, 108, 208

Delesseria sanguinea 62
Ecklonia sp. 108
Gelidium arbuscula 64
Gelidium canariensis 64
Gracilaria spp. 107
Hizakia fusiforme 62
Laurencia pinnatifida 65
Lomentaria articulata 62, 65, 107
Palmaria palmata 62, 65
Pterocladiella capillacea 64
Ulva lactuca 64, 109

Phytoplankton 1, 13, 29–32, 34, 41, 58, 60–62, 65, 68, 72, 101, 102, 109–111, 128, 138, 168, 169, 171, 172, 174, 182, 202–204, 222, 229, 230

Dunaliella tertiolecta 204
Monochrysis lutheri 204

Seagrasses 63, 64, 107, 108, 230

Cymodocea nodosa 63
Cymodocea serrulata 64
Halophila ovalis 64
Phyllospadix torreyi 63
Posidonia oceanica 63, 107

ANNELIDS 79, 126, 229

Nereis virens 126
Riftia pachyptila 79, 80
Sipunculus nudus 80, 127

BACTERIA 34, 57, 79, 80, 101, 182

Vibrio campbelli 133
Vibrio spp. 133
Vibrio tubiashii 124

BIRDS 88, 140, 148, 188, 230

Ducks 88

Aythya spp. 88

Penguins 88

Aptenodytes forsteri 88
Aptenodytes patagonicus 88

BRYOZOANS 139

Myriapora truncata 176

CHAETOGNATHS 140

Sagitta elegans 140

COELENTERATES 69, 100, 113

Anemones 74

Aiptasia pulchella 77
Anemonia viridis 77, 116

Corals 4, 5, 17, 35, 38, 51, 53, 55, 69–77, 100, 101, 114, 115, 117–119, 137, 141, 165, 170, 176, 184, 198, 206, 208, 218, 229, 230

Acropora cervicornis 209
Acropora eurystoma 69

Acropora intermedia 119
Acropora palmata 116
Acropora sp. 74, 76
Acropora verweyi 119
Astrangia danae 73
Corallium rubrum 71
Favia fragum 118
Favia sp. 76
Galaxea fascicularis 71, 73, 119
Lophelia pertusa 115
Madracis auretenra 73
Madracis mirabilis sensu 73
Madracis pharencis 117
Montipora capitata 117
Oculina patagonica 117
Pavona cactus 119
Pocillopora damicornis 117
Porites compressa 71, 119
Porites lobata 119
Porites porites 73
Porites sp. 198
Stylophora pistillata 71, 72, 116, 207
Tubastrea aurea 75
Turbinaria reniformis 119

CRUSTACEANS AND OTHER ARTHROPODS

Amphipods 129, 130, 218

Echinogammarus marinus 129
Eurythenes obesus 129
Gammarus locusta 130

Barnacles 130, 131

Amphibalanus amphitrite 130
Elminius modestus 131
Semibalanus balanoides 130, 131

Copepods 37, 40, 127–129, 131, 217

Acartia spp. 128
Acartia tsuensis 128
Calanus finmarchicus 128
Calanus helgolandicus 129
Calanus pacificus 127
Eucalanus bungii bungii 127
Metridia pacifica 127
Neocalanus cristatus 127
Neocalanus tonsus 37

Paraeuchaeta elongata 127
Pseudocalanus elongatus 129

Crabs 59, 81, 133, 134, 177

Callinectes sapidus 81, 134
Cancer magister 133
Cancer pagurus 133
Carcinus maenas 81, 134
Carcinus mediterraneus 81
Chionocetes bairdi 133
Eupagurus bernhardus 134
Hyas araneus 133
Limulus polyphemus 134
Necora puber 134

Euphausids 127, 131

Euphausia pacifica 127, 131
Euphausia superba 132

Lobsters 135

Homarus gammarus 134

Mysidaceans

Gnathophausia ingens 131

Ostracods 127

Conchoecia sp. 127

Shrimps and Prawns

Artemia franciscana 131
Glyphocrangon vicaria 82
Litopenaeus vannamei 133
Marsupenaeus japonicus 132
Metapenaeus macleayi 132
Palaemon pacificus 132
Palaemonetes pugio 133
Penaeus monodon 132
Penaeus occidentalis 132

ECHINODERMS 34, 37, 70, 100, 101, 121, 135, 137, 139, 165, 194, 229

Asteroids 200

Asterias rubens 139
Crossaster papposus 138
Patirella regularis 200
Pisaster ochraceus 139

Brittle stars 37

Amphiura filiformis 138
Ophiothrix fragilis 37, 137
Ophiura ophiura 201

Echinoids 38, 136, 194, 200

Centrostephanus rodgersii 200
Heliocidaris erythrogramma 137, 200, 201
Heliocidaris tuberculata 200
Tripneustes gratilla 135, 200

Sea urchins 36, 38, 100, 118, 128, 135–137, 139, 218

Arbacia punctata 136
Diadema savignyi 36
Echinometra mathaei 36, 136
Echinometra spp. 136
Echinocardium cordatum 139
Echinus esculentus 137
Evechinus chloroticus 135
Heliocidaris erythrogramma 137, 200, 201
Hemicentrotus pulcherrimus 136
Lytechinus pictus 136
Psammechinus miliaris 135, 137
Pseudechinus huttoni 135
Sterechinus neumayeri 135
Strongylocentrotus droebachiensis 136
Strongylocentrotus franciscanus 139
Strongylocentrotus purpuratus 137, 138

ELASMOBRANCHS 82, 83, 140, 229

Rays 140

Raja ocellata 83

Sharks 140

Mustelus manazo 143, 146
Scyliorhinus canicula 140
Scyliorhinus stellaris 82
Squalus acanthias 82

FISHES 83, 140

Amberjacks 143

Seriola lalandi 144
Seriola quinqueradiata 141, 142, 146

Antarctic notothenioids 87

Gobionotothen gibberifrons 87
Lepidonotothen kempi 147
Notothenia corticeps 87
Pachycara brachycephalum 147

Basses

Acanthopagrus australis 143
Dicentrarchus labrax 140, 141
Macquaria novemaculeata 143
Morone saxatilis 147
Pagrus aurata 143
Pagrus major 140–142
Sparus aurata 140, 146

Clownfish 142

Amphiprion percula 142

Croakers

Leiostomus xanthurus 146

Eelpouts 87

Zoarces viviparus 87

Eels 84, 87, 218

Anguilla anguilla 84
Conger conger 97

Flounders 84, 85, 146

Paralichthys olivaceus 84, 140, 146
Platichthys flesus 85
Pleuronectes platessa 141

Gadoids

Gadus morhua 145

Herrings 141

Clupea harengus 141

Killifishes

Fundulus heteroclitus 84, 146, 147

Parrotfishes 218

Chlorurus sordidus 36

Salmons 144

Salmo salars 144

Smelts 141

Atherina boyeri 141

Sturgeons 85

Acipenser transmontanus 85, 143

Toadfishes 39, 86

Opsanus beta 39

Tunas 140

Euthynnus affinis 140

Whitings 140, 144

Sillago japonica 140, 144

Wolffishes 144

Anarhichas minor 144

MAMMALS 88, 140, 149, 188, 230

Dolphins 88

Tursiops truncatus 88

Seals 88, 89, 149

Callorhinus spp. 88
Halichoerus grypus 89
Leptonychotes weddelli 88
Mirounga angustirostris 88, 149
Mirounga leonina 149

MOLLUSCS 34–37, 53, 59, 72, 78, 100, 101, 120–122, 124, 125, 128, 141, 148, 177, 229, 230

Cephalopods 59

Dosidecus gigas 201
Sepia lycidas 125
Sepia officinalis 125
Sepioteuthis lessoniana 125

Clams 78, 79, 120, 123, 126

Calyptogena magnifica 79
Mercenaria mercenaria 78, 120, 123
Tindaria callistiformis 79
Venerupis decussata 120
Venerupis pallustra 126

Gastropods 70, 100, 124

Haliotis coccoradiata 200, 201
Haliotis laevigata 124
Haliotis rubra 125
Haliotis rufescens 125
Littorina littorea 124
Littorina obtusata 124
Limacina helicina 36, 124, 186
Thais gradata 173

Mussels 120, 122, 123, 177

Bathymodiolus brevior 177
Mytilus edulis 120, 122, 123
Mytilus galloprovincialis 122

Oysters 120–126

Crassostrea ariakensis 121
Crassostrea gigas 78, 121, 123
Crassostrea virginica 120–123, 125
Ostrea edulis 120

Pinctada fucata 121
Saccostrea glomerata 121

Scallops 123, 126

Argopecten irradians 123

NEMERTEANS 126

Procephalothrix simulus 126

REPTILES 87, 140, 147, 148, 230

Turtles 87, 147, 148

Caretta caretta 148
Chelonia mydas 87, 147

SPONGES 182

Aphrocallistes vastus 182
Rhadocalyptus dawsoni 182

TUNICATES 34, 38

Bathochordaeus sp. 38

About the Author

Ronald Eisler received the B.A. degree from New York University in biology and chemistry, and the M.S. and Ph.D. degrees from the University of Washington in aquatic sciences and radioecology, respectively. He has conducted research at laboratories of the University of Miami at Coral Gables, Florida, the University of Washington at Seattle, the Hebrew University of Jerusalem in Eilat, Israel, the U.S. Department of the Army Medical Service Corps in Denver, Colorado, the U.S. Department of the Interior in Highlands, New Jersey and Laurel, Maryland, and the U.S. Environmental Protection Agency in Narragansett, Rhode Island. He has participated in research and monitoring studies in the Pacific Northwest, the Territory of Alaska, Colorado, the Marshall and Marianas Islands, all along the eastern seaboard of the United States Atlantic coast, the Adirondacks region of New York, the Gulf of Aqaba in the Red Sea, and the Gulf of Mexico. He is the author of more than 150 technical articles including several books and 16 book chapters—mainly on contaminant hazards to plants and animals. He has held a number of adjunct professor appointments and taught for extended periods at the Graduate School of Oceanography of the University of Rhode Island, and the Department of Biology of American University in Washington, D.C. He also served as Visiting Professor and Resident Director of Hebrew University's Marine Biology Laboratory in Eilat, Israel, as senior science advisor to the American Fisheries Society (Bethesda, Maryland), and as Acid Rain Coordinator for the U.S. Department of the Interior (Washington, D.C.). In retirement, he consults and writes on chemical risk assessment.

Eisler resides in Potomac, Maryland, with his wife, Jeannette, a retired teacher of French and Spanish.

ALSO BY RONALD EISLER

2010. *Compendium of Trace Metals and Marine Biota. Volume 1: Plants and Invertebrates,* Elsevier, Amsterdam. 610 pp.

2010. *Compendium of Trace Metals and Marine Biota. Volume 2: Vertebrates,* Elsevier, Amsterdam. 500 pp.

2007. *Eisler's Encyclopedia of Environmentally Hazardous Priority Chemicals,* Elsevier, Amsterdam. 950 pp.

2006. *Mercury Hazards to Living Organisms,* CRC Press, Boca Raton, Florida. 312 pp.

2004. *Biogeochemical, Health, and Ecotoxicological Perspectives on Gold and Gold Mining,* CRC Press, Boca Raton, Florida. 355 pp.

2000. *Handbook of Chemical Risk Assessment. Health Hazards to Humans, Plants, and Animals, Volume 1, Metals,* Lewis Publishers, Boca Raton, Florida. 738 pp.

2000. *Handbook of Chemical Risk Assessment. Health hazards to Humans, Plants, and Animals. Volume 2. Organics,* Lewis Publishers, Boca Raton, Florida. 762 pp.

2000. *Handbook of Chemical Risk Assessment. Health Hazards to Humans, Plants, and Animals. Volume 3, Metalloids, Radiation, Cumulative Index to Chemicals and Species,* Lewis Publishers, Boca Raton, Florida. 403 pp.

1981. *Trace Metal Concentrations in Marine Organisms,* Pergamon Press, Elmsford, New York. 687 pp.